PROGRESS IN ENVIRONMENTAL ENGINEERING

Progress in Environmental Engineering

Water, Wastewater Treatment and Environmental Protection Issues

Editors

Janusz A. Tomaszek & Piotr Koszelnik

*Department of Environmental & Chemistry Engineering,
Rzeszów University of Technology, Rzeszów, Poland*

CRC Press
Taylor & Francis Group
Boca Raton   London   New York   Leiden

CRC Press is an imprint of the
Taylor & Francis Group, an **informa** business

A BALKEMA BOOK

CRC Press/Balkema is an imprint of the Taylor & Francis Group, an informa business

© 2015 Taylor & Francis Group, London, UK

Typeset by MPS Limited, Chennai, India
Printed and bound in Great Britain by CPI Group (UK) Ltd, Croydon, CR0 4YY

Published by: CRC Press/Balkema
                P.O. Box 11320, 2301 EH Leiden, The Netherlands
                e-mail: Pub.NL@taylorandfrancis.com
                www.crcpress.com – www.taylorandfrancis.com

ISBN: 978-1-138-02799-2 (Hbk)
ISBN: 978-1-315-68547-2 (eBook PDF)

Progress in Environmental Engineering – Tomaszek & Koszelnik (eds)
© 2015 Taylor & Francis Group, London, ISBN: 978-1-138-02799-2

Table of contents

Progress in Environmental Engineering – Tomaszek & Koszelnik (eds)
© 2015 Taylor & Francis Group, London, ISBN: 978-1-138-02799-2

Preface

The monograph contains original theoretical and experimental papers dealing with: water purification, especially on risk management in water distribution system operation and maintenance, new concepts and methods of wastewater treatment e.g. elimination of activated sludge bulking or using a new support material in activated sludge technology, greenhouse gases emissions from WWTPs, and important ecological problems in freshwater ecosystems.

There have been many advances in the study of aquatic ecosystems in recent years, but there remain many questions to be solved. The areas that require new approach, in spite of the advances during the last decades, are the paramount eutrophical problems related to lakes and reservoirs restoration, the role of wetlands in the removal of heavy metals and complicated interactions between sediment and overlying water. This monograph contains contributions pointing to these directions. The goal of the monograph is not merely to provide technical proficiency but to add insight and understanding of the selected aspects of water purification, wastewater treatment and protection of aquatic ecosystems. We hope that the present monograph, by bringing together a plenty of information on origin, nature and reduction of environment contaminations, will help with providing modes of action to effectively solve the pollution problems.

The editors would like to express their acknowledgement to all the authors of the monograph for their enthusiasm, diligence and involvment.

We extend our gratitude to all those who helped with making the monograph.

Janusz A. Tomaszek and Piotr Koszelnik

About the editors

Janusz A. Tomaszek – Professor
Department of Environmental & Chemistry Engineering,
Rzeszów University of Technology, Poland

Professor, 2007, Environmental Engineering & Chemistry Engineering,
Warsaw University of Technology, Poland

Ph.D.Sc., 1992, Environmental Engineering & Chemistry Engineering,
Warsaw University of Technology, Poland

Ph.D., 1980, Polish Academy of Science, Zabrze, Silesia, Poland

Research Interests:
– *Water Chemistry/Ecosystem Dynamics:* transformations of organic compounds and nutrients, geochemistry of sediments, chemical processes at sediment-water interface, IRMS measurements, trace elements, heavy metals, GHG emissions.
– Water purification and sewage treatment.
– Water pollution control.

Piotr Koszelnik – Associate Professor
Department of Environmental & Chemistry Engineering,
Rzeszów University of Technology, Poland

Ph.D.Sc., 2009, Environmental Engineering, Environmental
Chemistry, Warsaw University of Technology, Poland

Ph.D., 2003, Environmental Engineering, Lublin University of
Technology, Poland

Research Interests:
– Environmental chemistry especially water chemistry: eutrophication, carbon and nitrogen cycling, stable isotopes, reclamation of man-made lakes, micropollutants in water.
– Waste management and utilization.

Progress in Environmental Engineering – Tomaszek & Koszelnik (eds)
© *2015 Taylor & Francis Group, London, ISBN: 978-1-138-02799-2*

Risk management in water distribution system operation and maintenance using Bayesian theory

B. Tchórzewska-Cieślak & K. Pietrucha-Urbanik
Rzeszów University of Technology, Rzeszów, Poland

ABSTRACT: Water Distribution System (WDS) is one of the basic technological systems belonging to the underground infrastructure which has a priority importance for people's lives. In water distribution system operation we deal with events that can cause breaks in water supply or water pollution. For purposes of this paper operational reliability of the WDS is defined as the ability to supply a constant flow of water for various groups of consumers, with a specific quality and a specific pressure, according to consumers demands, in the specific operational conditions, at any or at specific time and safety of the WDS means the ability of the system to safely execute its functions in a given environment. The main aim of this paper is to present a method for the risk management using Bayesian process. The proposed method made it possible to estimate the risks associated with the possibility of partial or total loss of the ability of water supply system operation.

1 INTRODUCTION

Risk management of failures of Water Distribution System (WDS) is a set of organizations, institutions, technical systems, education and control, which aim is to ensure the safety of water consumers. The management system is introduced on the level of the local water companies. Risk management is part of a modern and well-developed system of safety management of water supply systems. It is a multi-step procedure aimed at improving the system safety, including quantitative and qualitative aspects of drinking water (Tchórzewska-Cieślak 2011). This process is based primarily on the risk analysis, risk assessment or risk estimation, making decision on its acceptability, periodic control or reduction (Hastak & Baim 2001, Walkowiak & Mazurkiewicz 2009). Risk as a measure of loss of WDS safety associated with the production and distribution of drinking water, refers to the likelihood of undesirable events and the size of potential losses and vulnerability to threat (or the degree of protection) (Juraszka & Braun 2011, Kruszynski & Dzienis 2008, Li et al. 2009, Pollard et al. 2004, Rak & Pietrucha 2008, Valis et al. 2010). Risk management should be considered as a process inseparably linked to the management of the whole water supply company by developing methods for response to risk, that means preparing the organizational infrastructure supporting risk management (Tchórzewska-Cieślak 2007, Tchorzewska-Cieślak & Rak 2010). Risk identification is based on a selection of representative emergency events that may occur during continuous operation of WDS, including initiating events that could cause the so-called domino effect (Rak 2009). Risk assessment is the process of its qualitative and quantitative analysis, using adequate for the type of risk methods, with determining the criterion value for the adopted scale of risk, for example, the three-stage scale, which distinguishes tolerated, controlled and unacceptable risk (Apostolakis & Kaplan 1981, Boryczko & Tchórzewska-Cieślak 2013, Tchórzewska-Cieślak & Kalda 2008). Due to the large complexity of the individual elements of the system and their spatial dispersion, diverse methods of risk assessment are applied (Mazurkiewicz & Walkowiak 2004, Studzinski & Pietrucha-Urbanik 2012).

Generally WDS includes the water supply network (main and distribution) with fittings, tanks and pumping stations.

It is clear that during WDS operation the different types of failure, which may cause loss of water as well as a break in water supply and the so-called secondary water contamination in the water supply network, can appear (Rak & Pietrucha 2008, Pietrucha-Urbanik & Tchórzewska-Cieślak 2013).

Threats to the whole WDS can be classified according to the type of cause:

– internal (resulting directly from the operation of the system, such as damage to its components, failure in main or distribution pipes and fittings, pumping stations failures),
– external (e.g. incidental pollution of water source, forces of nature, such as flood, drought, heavy rains, storms, landslides, as well as the lack of power supply or actions of third parties (e.g. vandalism, terrorist attack, cyber-terrorist attack).

The most common undesirable events in WDS are failures of water supply pipes and fittings. In most cases failures of fittings are not a direct threat to water consumers. It also applies to water leaks in pipes that do not cause the need to exclude the network segment from the operation (Christodoulous 2008, Studzinski & Pietrucha-Urbanik 2012). Due to the specificity of water supply system operation the failure removal is inseparably connected with the maintaining the network reliability and the priority is to provide consumers with water of appropriate quality, at the right pressure, at any time.

2 RISK ANALYSIS

Loss of WDS safety always causes a risk of negative consequences felt by water consumers. It is associated with:

– lack or interruption in water supply,
– health threat for water consumers as a result of consuming poor quality drinking water,
– consumers financial losses, for example, the need to purchase bottled water, treatment costs, costs arising from the hygienic and sanitary difficulties.
– Consumer's risk is a function of the following parameters:
– a measure of the probability P or the frequency of the occurrence of undesirable events in WDS which are directly felt by water consumers,
– losses C associated with it (e.g. purchase of bottled water, any medical expenses after consuming unfit for drinking water or immeasurable losses, such as living and economic difficulties or loss of life or health),
– the degree of vulnerability to undesirable events V or the degree of protection against undesirable events O.

Consumer's risk (individual) r_K is the sum of the first kind risk r_{KI}, associated with the possibility of interruptions in water supply, and the second kind risk r_{KII}, associated with the consumption of poor quality water (Tchórzewska-Cieślak 2011).

For the risk of the first type, the three parametric definition was assumed:

$$r_{KI} = \sum_{RSA=1}^{N_I} (f_{iI} \cdot C_{jI} \cdot V_{kI}) \tag{1}$$

where RSA = sequence of consecutive undesirable events (or a single undesirable event) that may cause the risk of the first type; I = adopted scale for the frequency parameter; j = adopted scale for the loss parameter, k = adopted scale for the vulnerability parameter; f_{iI} = frequency (or likelihood) of the RSA occurrence or a single event that may cause the risk of the first type; C_{jI} = losses caused by the given RSA or a single undesirable event that may cause the risk of the first type; V_{kI} = vulnerability associated with the occurrence of the given RSA or a single undesirable event that may cause the risk of the first type; N_I = number of RSA or individual undesirable events; and N_I = number of RSA or single undesirable events.

For the consumer risk of the second type, the following definition was assumed:

$$r_{KII} = \sum_{RSA=1}^{N_I} (f_{iII} \cdot C_{jII} \cdot V_{kII}) \tag{2}$$

where RSA = sequence of consecutive undesirable events (or a single undesirable event) that may cause the risk of the second type; f_{iII} = frequency (likelihood) of the RSA occurrence or a single event that may cause the risk of the second type; C_{jII} = the value of losses connected with health threat caused by given RSA or a single undesirable event that may cause the second kind risk; V_{kII} = vulnerability to the occurrence of RSA or a single failure event that may cause the second kind risk; and N_{II} is a number of RSA or single undesirable events.

The risk analysis for the WDS safe operation should be conducted in the following stages of reconnaissance:

– determining the number of people using the WDS,
– determining the representative failure events and analysing their crisis scenarios in order to estimate losses,
– determining the probability (frequency) of undesirable events,
– determining the vulnerability degree of water consumers to undesirable events
– analysing the WDS protection system, including system monitoring and remote control, and the so called protective barriers included in the WDS, for example, alternative water intakes or multi-barrier systems (Rak 2009),
– estimating potential losses, including the probability of exceeding a certain value of limit losses,
– determining the risk level in the five-stage scale.

3 THE USE OF BAYESIAN MODELS IN RISK ANALYSIS

3.1 *Scope of the data and measurements needed for WDS risk analysis*

Indicators and measures that can be used in the process of WDS risk analysis generally are divided into:

– statistical – determined in accordance with accepted principles of mathematical statistics based on historical data from the operation of the subsystem,
– probabilistic – determined on the basis of the probability theory,
– linguistic – describing the risk parameters by means of the so-called linguistic variables, expressed in natural language by such words as: small, medium, large.

Key indicators, measures and functions used to estimate the individual risk parameters are (Kwietniewski et al. 1993, Tchórzewska-Cieślak 2011):

– n_a – a number of failures during the analysed period of WDS operation,
– n_{aj} – a number of failures (undesirable events) caused by a specific factor j for the analysed period of WDS operation,
– n_{ai} – a number of failures (undesirable events) that cause a specific effect i for the analysed period of WDS operation,
– the average values of the number of undesirable events (failures) together with the basic statistical characteristics, such as median, standard deviation, lower and upper quartile, the degree of dispersion,
– the average operating time between failures T_p [d], which is the expected value of a random variable T_p defining operating time (ability of the system (or its components) between two consecutive failures,
– the mean repair time T_n [h] is interpreted as the expected value of time from a moment of failure to a moment when an element is included to the operation. It is the sum of the waiting for repair

time T_d and the real repair time T_0 (till the inclusion of the element to the operation):

$$T_n = T_d + T_0 \qquad (3)$$

The analysis of the WDS operation in terms of water consumers safety must also take into account as a component of failure repair time, the time of interruptions in water supply to customers.

The failure rate $\lambda(t)$ [number of failures $\cdot$ year (day)$^{-1}$] or [number of failures $\cdot$ km^{-1}a^{-1}] is calculated according to the formulas (Kwietniewski et al. 1993):

$$\lambda = \frac{1}{T_p} \qquad (4)$$

and for linear elements:

$$\lambda(t) = \frac{n(t, t+\Delta t)}{N \cdot \Delta t} \qquad (5)$$

where T_p = the average time between subsequent failures; $n(t, t + \Delta t)$ = total number of failures in the time interval $(t, t + \Delta t)$; N = number of analysed elements or for linear elements their length L [km]; and Δt = time of observation.
– the repair rate $\mu(t)$ [number of repairs$\cdot$a(h)$^{-1}$] determines the number of failures repaired per time unit, it can be determined from the operating data according to the formula (with assumption of Poison stream of failures):

$$\mu = \frac{1}{T_n} \qquad (6)$$

– the frequency of failures f is calculated as the average number of failures (damages, undesirable events) per time unit during the operation [failure/s, failure/month].

3.2 *Principles of Bayesian data classification*

Random nature of the formation of failure causes that related to it research is complex and is based primarily on the analysis of operational data and experts opinions. The idea of data exploration involves the use of information technology to find information in databases. There are many data exploration techniques derived directly from mathematical statistics and machine learning (Bishop 2006, Zitrou et al. 2010, Zhang & Horigome 2001).

The task of classification is to create a model that allows you to assign an unknown element or its attribute to a predefined set (class). It consists in the construction of decision rule to classify observations as realizations of particular classes of objects' similarity. Classification methods (Larose 2006, Morzy 2007):

– pattern recognition – used when you have some information about the classes from which information was taken (e.g. discriminant analysis),
– no pattern recognition – used when the analysed sample contains not classified observations or those that cannot be used to build the classification functions.

All the methods of classification should be characterized by:

– unambiguity – one element can belong to one class only,
– transparency of the classification rules,
– the ability to modify the classification rules.

An important issue in the classification process is the selection of diagnostic variables. The basis of this selection is to develop a preliminary list of the characteristics of the analysed objects (e.g. water mains, pumping stations). Diagnostic variables should be (Morzy 2007, Ritter & Gallegos 2002):

– weakly correlated or uncorrelated,
– strongly correlated with variables that are not in the diagnostic team and should not be influenced externally.

Elements e_i that are subject to classification create the set Ω, where $\Omega = \{e_1, \ldots, e_n\}$, while a set of characteristics x_j is adopted to describe the classified elements due to the studied phenomenon – is implemented by a set of random variables $X = \{x_1, \ldots, x_k\}$, with probability density $f(x_k)$. Variables x_{ij} are called the diagnostic variables. The data matrix M_d is written in the following way:

$$M_d = \left[\overrightarrow{x_{ij}} \right] = \begin{bmatrix} x_{11}, \ldots, x_{n1} \\ x_{1k}, \ldots, x_{nk} \end{bmatrix} \qquad (7)$$

where x_{ij} = diagnostic variable; $i = 1, 2, \ldots, n$; n is the number of elements of the set Ω; $j = 1, 2, \ldots, k$; and k = number of features considered in the classification.

Lines characterize elements i and columns features j. The matrix is called the data matrix in which each element e_i is characterized by the vector x_{ij}.

A classifier d (a classification rule) is the function $F(X)$, which assigns to each x_{ij} the specific class of a given set of classes: $d: X \rightarrow KL_l = \{1, 2, \ldots, l\}$, where l is the number of class and d is a classification rule.

The basic principles of Bayesian classification (Bernardo & Smith 1993, Bishop 2006, Ritter & Gallegos 2002, Tchórzewska-Cieślak 2011):

– for l different classes: $KL_A, KL_B, \ldots, KL_L$, the Bayes theorem is in form (Bernardo & Smith 1993, Bishop 2006):

$$P(KL_A \mid x_{ij}) = \frac{P(x_{ij} \mid KL_A) \cdot P(KL_A)}{p(x_{ij})} \qquad (8)$$

$$p(x_{ij}) = \sum_{A=1}^{l} p(x_{ij} \mid KL_A) \cdot P(KL_A) \qquad (9)$$

$$\sum_{A=1}^{l} P(KL_A \mid x_{ij}) = 1 \qquad (10)$$

where KL_L = class designation; l is number of classes; $P(KL_A)$ is a priori probability for class A; $P(x_{ij}/KL_A)$ is likelihood, reliability that the element is described by the vector x_{ij} and class A occurs; $P(KL_A/x_{ij})$ – a posteriori probability of the hypothesis that element x_{ij} belongs to class KL_A; $p(x_{ij})$ is density of probability of x_{ij} occurrence, the so-called total evidence, the scaling factor,

– rule d includes x_{ij} to class A, if $x_{ij} \in KL_A$.
– $C(KL_B/KL_A)$ means a loss caused by classifying x_{ij} into class B while in reality it belongs to class A, $0 < C(KL_B/KL_A) < \infty$, $KL_A \neq KL_B$, $A, B = 1, \ldots, l$,
– the probability of erroneous classification of x_{ij} is defined by the relation:

$$P(KL_B \mid KL_A) = \int_X f(x_k) dx \qquad (11)$$

– risk $r_A(d)$ of erroneous classification is given by the formula:

$$r_A(d) = \sum_{A,B=1}^{L} C(KL_B \mid KL_A) \cdot P(KL_B \mid KL_A), \quad A, B = 1, \ldots, l \quad A \neq B \qquad (12)$$

- the risk set $r_A(d) = \{r_1(d), \ldots, r_L(d)\}$ characterizes a classification rule d,
- if there are two classification rules: d_1, d_2 and $r_1(d), r_2(d)$, then rule d_1 is more favourable than $d_2, r_A(d_1) \leq r_A(d_2), A = 1, 2, \ldots, l$ and when at least for one feature j the condition $r_A(d_1) < r_A(d_2)$ is fulfilled. If for all the features $r_A(d_1) = r_A(d_2)$, then both rules are equivalent,
- if $r_A(d_1) > r_A(d_2)$, then the rules are not comparable, until new criteria are introduced, classification rule is optimal (acceptable), if there is no more favourable rule,
- when the probability density distribution is known a priori for the fact that classified x_{ij} belongs to class A, $p(KL_A)$, the absolute value of the expected loss corresponding to the classification rule d is called the Bayesian risk r_B:

$$r_B(d) = \sum_{A=1}^{L} r_A(d) \cdot p(KL_A) = \sum_{A=1X}^{L} \int u(x)dx \tag{13}$$

where $u(x)$ – the classification function is:

$$u(x) = -\sum_{\alpha=1}^{L} p(\alpha) \cdot C(K_A | KL_\alpha) \cdot f_\alpha(x) \tag{14}$$

In order to minimize the losses, element e_i must be assigned to the class for which it is the smallest.

- For a simple loss function:

$$C(KL_B | KL_A) = \begin{cases} 1, & KL_A \neq K_B \\ 0, & KL_A = K_B \end{cases} \tag{15}$$

- The Bayesian risk r_B is given by:

$$r_B(d) = 1 - \sum_{A=1}^{L} p(KL_A) \cdot C\left(KL_B | KL_A\right) \tag{16}$$

- The classification function $u(x)$ takes the form:

$$u(x) = p(KL_A) \cdot f_A(x) \tag{17}$$

A classification rule d is the Bayesian against a priori distribution $P(KL_A)$, if it minimizes the Bayesian risk.

3.3 Risk model using the Bayesian network

The Bayesian networks – BRA (Bayes Risk Analysis) are used in risk analysis due to the ability to model the dependent events. The Bayesian network is upgraded by means of experience and acquired knowledge. The network is modelled by a directed acyclic graph in which vertices represent events and edges represent causal connections between these events. In addition, the Bayesian network is not limited to two states: up state or down state (as in the event tree method and the fault tree method) and may be used for analysing the intermediate states.

The relations between the vertices (events) are expressed by means of the conditional probability. For the vertex X, whose parents are in the set $\pi(X)$, these relations are represented by the conditional probability tables (CPT). In CPT, for the variable X, all the probabilities $P(X|\pi(X))$ (for all the possible combinations of variables from the set $\pi(X)$) must be specified. The table for the vertex that does not have parents includes the probabilities that the random variable X will take its particular values.

6

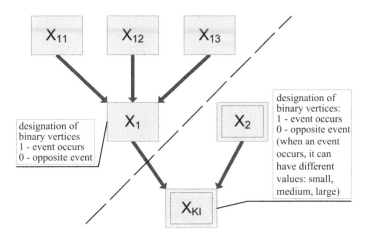

Figure 1.   Bayesian network for the risk of the first type.

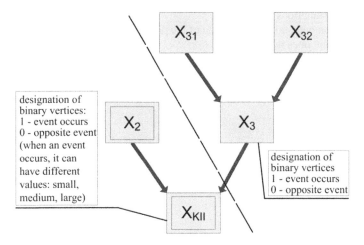

Figure 2.   Bayesian network for the risk of the second type.

If the network has n vertices, $X_1, \ldots, X_n$, the total probability distribution of all the random variables is shown as the relation (Bishop 2006, Tchórzewska-Cieślak 2010, Tchórzewska-Cieślak & Włoch 2006):

$$P(X_1, \ldots, X_n) = \prod_{n=1}^{n} P(X_i | \pi(X_i))$$

(18)

The Bayesian network can be used in the decision-making model analysing the risk of failure in water distribution subsystem (Tchórzewska-Cieślak & Włoch 2006).

In Figures 1 and 2 the developed Bayesian network schemes, used for failure risk analysis of water distribution subsystem, from the water consumer point of view, are presented.

Symbols used in Figure 1 and 2 mean (Tchórzewska-Cieślak 2013):

$r_{KI,II}$ – consumer's risk (the first or second type) in point scale:

– tolerable risk: $r_{KI,II} = r_{K1}$,
– controlled: $r_{KI,II} = r_{K2}$,
– unacceptable: $r_{KI,II} = r_{K3}$,

X_1 – interruption in water supply

- X_{11} – failure of the water supply network,
- X_{12} – lack of water supply from the water treatment plant,
- X_{13} – failure of zone pumping stations,

X_2 – consumers protection from the existing threat

- very little – x_{21},
- little – x_{22},
- medium – x_{23},
- large – x_{24},
- very large – x_{25},

X_3 – water quality parameters specified in the relevant Regulations of the Minister of Health are exceeded,

- X_{31} – physico-chemical parameters are exceeded,
- X_{32} – microbiological parameters are exceeded.

The following assumptions were made:
the event in the given node takes exactly one of the possible values,1 means that the event occurs, 0 means that the event does not occur.

For each vertex the CPT should be defined (Tchórzewska-Cieślak 2013):
For the risk of the first kind:

- $P(r_{KI}|X_1, X_2)$,
- $P(X_1|X_{11}, X_{12}, X_{13})$,
- $P(X_2)$,
- $P(X_{11})$,
- $P(X_{12})$,
- $P(X_{13})$,

For the risk of the second kind

- $P(r_{KII}|X_2, X_3)$,
- $P(X_3|X_{31}, X_{32})$,
- $P(X_{31})$,
- $P(X_{32})$.

The probability that the consumer's risk of the first type is (according the equations 9–10, 18):
For r_{KI} the particular probability values are:

- the probability that the consumer's risk of the first kind is tolerable:
$P(r_{KI} = r_{KI1}) =$
$P(r_{KI}=r_{KI1}|X_1=1 \wedge X_2=x_{21}) \cdot P(X_1=1) \cdot P(X_2=x_{21}) + P(r_{KI}=r_{KI1}|X_1=1 \wedge X_2=x_{22}) \cdot P(X_1=1) \cdot P(X_2=x_{22}) +$
$P(r_{KI}=r_{KI1}|X_1=1 \wedge X_2=x_{23}) \cdot P(X_1=1) \cdot P(X_2=x_{23}) + P(r_{KI}=r_{KI1}|X_1=1 \wedge X_2=x_{24}) \cdot P(X_1=1) \cdot P(X_2=x_{24}) +$
$P(r_{KI}=r_{KI1}|X_1=1 \wedge X_2=x_{25}) \cdot P(X_1=1) \cdot P(X_2=x_{25}) + P(r_{KI}=r_{KI1}|X_1=0 \wedge X_2=x_{21}) \cdot P(X_1=0) \cdot P(X_2=x_{21}) +$
$P(r_{KI}=r_{KI1}|X_1=0 \wedge X_2=x_{22}) \cdot P(X_1=0) \cdot P(X_2=x_{22}) + P(r_{KI}=r_{KI1}|X_1=0 \wedge X_2=x_{23}) \cdot P(X_1=0) \cdot P(X_2=x_{23}) +$
$P(r_{KI}=r_{KI1}|X_1=0 \wedge X_2=x_{24}) \cdot P(X_1=0) \cdot P(X_2=x_{24}) + P(r_{KI}=r_{KI1}|X_1=0 \wedge X_2=x_{25}) \cdot P(X_1=0) \cdot P(X_2=x_{25})$.
- the probability that the consumer's risk of the first kind is controlled:
$P(r_{KI}=r_{KI2}) =$
$P(r_{KI}=r_{KI2}|X_1=1 \wedge X_2=x_{21}) \cdot P(X_1=1) \cdot P(X_2=x_{21}) + P(r_{KI}=r_{KI2}|X_1=1 \wedge X_2=x_{22}) \cdot P(X_1=1) \cdot P(X_2=x_{22}) +$
$P(r_{KI}=r_{KI2}|X_1=1 \wedge X_2=x_{23}) \cdot P(X_1=1) \cdot P(X_2=x_{23}) + P(r_{KI}=r_{KI2}|X_1=1 \wedge X_2=x_{24}) \cdot P(X_1=1) \cdot P(X_2=x_{24}) +$
$P(r_{KI}=r_{KI2}|X_1=1 \wedge X_2=x_{25}) \cdot P(X_1=1) \cdot P(X_2=x_{25}) + P(r_{KI}=r_{KI2}|X_1=0 \wedge X_2=x_{21}) \cdot P(X_1=0) \cdot P(X_2=x_{21}) +$
$P(r_{KI}=r_{KI2}|X_1=0 \wedge X_2=x_{22}) \cdot P(X_1=0) \cdot P(X_2=x_{22}) + P(r_{KI}=r_{KI2}|X_1=0 \wedge X_2=x_{23}) \cdot P(X_1=0) \cdot P(X_2=x_{23}) +$
$P(r_{KI}=r_{KI2}|X_1=0 \wedge X_2=x_{24}) \cdot P(X_1=0) \cdot P(X_2=x_{24}) + P(r_{KI}=r_{KI2}|X_1=0 \wedge X_2=x_{25}) \cdot P(X_1=0) \cdot P(X_2=x_{25})$.
- the probability that the consumer's risk of the first kind is unacceptable:
$P(r_{KI}=r_{KI3}) =$
$P(r_{KI}=r_{KI3}|X_1=1 \wedge X_2=x_{21}) \cdot P(X_1=1) \cdot P(X_2=x_{21}) + P(r_{KI}=r_{KI3}|X_1=1 \wedge X_2=x_{22}) \cdot P(X_1=1) \cdot P(X_2=x_{22}) +$

$P(r_{KI}=r_{KI3}|X_1=1 \wedge X_2=x_{23}) \cdot P(X_1=1) \cdot P(X_2=x_{23}) + P(r_{KI}=r_{KI3}|X_1=1 \wedge X_2=x_{24}) \cdot P(X_1=1) \cdot P(X_2=x_{24}) +$
$P(r_{KI}=r_{KI3}|X_1=1 \wedge X_2=x_{25}) \cdot P(X_1=1) \cdot P(X_2=x_{25}) + P(r_{KI}=r_{KI3}|X_1=0 \wedge X_2=x_{21}) \cdot P(X_1=0) \cdot P(X_2=x_{21}) +$
$P(r_{KI}=r_{KI3}|X_1=0 \wedge X_2=x_{22}) \cdot P(X_1=0) \cdot P(X_2=x_{22}) + P(r_{KI}=r_{KI3}|X_1=0 \wedge X_2=x_{23}) \cdot P(X_1=0) \cdot P(X_2=x_{23}) +$
$P(r_{KI}=r_{KI3}|X_1=0 \wedge X_2=x_{24}) \cdot P(X_1=0) \cdot P(X_2=x_{24}) + P(r_{KI}=r_{KI3}|X_1=0 \wedge X_2=x_{25}) \cdot P(X_1=0) \cdot P(X_2=x_{25}).$

For r_{KII} the particular values of the probabilities were determined in the same way as for the probability of the first kind:

– the probability that the consumer's risk of the second kind is tolerable:
$P(r_{KII}=r_{KII1}) =$
$P(r_{KII}=r_{KII1}|X_2=x_{21} \wedge X_3=1) \cdot P(X_2=x_{21}) \cdot P(X_3=1) + P(r_{KII}=r_{KII1}|X_2=x_{22} \wedge X_3=1) \cdot P(X_2=x_{22}) \cdot P(X_3=1) +$
$P(r_{KII}=r_{KII1}|X_2=x_{23} \wedge X_3=1) \cdot P(X_2=x_{23}) \cdot P(X_3=1) + P(r_{KII}=r_{KII1}|X_2=x_{24} \wedge X_3=1) \cdot P(X_2=x_{24}) \cdot P(X_3=1) +$
$P(r_{KII}=r_{KII1}|X_2=x_{25} \wedge X_3=1) \cdot P(X_2=x_{25}) \cdot P(X_3=1) + P(r_{KII}=r_{KII1}|X_2=x_{21} \wedge X_3=0) \cdot P(X_2=x_{21}) \cdot P(X_3=0) +$
$P(r_{KII}=r_{KII1}|X_2=x_{22} \wedge X_3=0) \cdot P(X_2=x_{22}) \cdot P(X_3=0) + P(r_{KII}=r_{KII1}|X_2=x_{23} \wedge X_3=0) \cdot P(X_2=x_{23}) \cdot P(X_3=0) +$
$P(r_{KII}=r_{KII1}|X_2=x_{24} \wedge X_3=0) \cdot P(X_2=x_{24}) \cdot P(X_3=0) + P(r_{KII}=r_{KII1}|X_2=x_{25} \wedge X_3=0) \cdot P(X_2=x_{25}) \cdot P(X_3=0).$
– the probability that the consumer's risk of the second kind is controlled:
$P(r_{KII}=r_{KII2}) =$
$P(r_{KII}=r_{KII2}|X_2=x_{21} \wedge X_3=1) \cdot P(X_2=x_{21}) \cdot P(X_2=1) + P(r_{KII}=r_{KII2}|X_2=x_{22} \wedge X_3=1) \cdot P(X_2=x_{22}) \cdot P(X_3=1) +$
$P(r_{KII}=r_{KII2}|X_2=x_{23} \wedge X_3=1) \cdot P(X_2=x_{23}) \cdot P(X_3=1) + P(r_{KII}=r_{KII2}|X_2=x_{24} \wedge X_3=1) \cdot P(X_2=x_{24}) \cdot P(X_3=1) +$
$P(r_{KII}=r_{KII2}|X_2=x_{25} \wedge X_3=1) \cdot P(X_2=x_{25}) \cdot P(X_3=1) + P(r_{KII}=r_{KII2}|X_2=x_{21} \wedge X_3=0) \cdot P(X_2=x_{21}) \cdot P(X_3=0) +$
$P(r_{KII}=r_{KII2}|X_2=x_{22} \wedge X_3=0) \cdot P(X_2=x_{22}) \cdot P(X_3=0) + P(r_{KII}=r_{KII2}|X_2=x_{23} \wedge X_3=0) \cdot P(X_2=x_{23}) \cdot P(X_3=0) +$
$P(r_{KII}=r_{KII2}|X_2=x_{24} \wedge X_3=0) \cdot P(X_2=x_{24}) \cdot P(X_3=0) + P(r_{KII}=r_{KII2}|X_2=x_{25} \wedge X_3=0) \cdot P(X_2=x_{25}) \cdot P(X_3=0).$
– the probability that the consumer's risk of the second kind is unacceptable:
$P(r_{KII}=r_{KII3}) =$
$P(r_{KII}=r_{KII3}|X_2=x_{21} \wedge X_3=1) \cdot P(X_2=x_{21}) \cdot P(X_3=1) + P(r_{KII}=r_{KII3}|X_2=x_{22} \wedge X_3=1) \cdot P(X_2=x_{22}) \cdot P(X_3=1) +$
$P(r_{KII}=r_{KII3}|X_2=x_{23} \wedge X_3=1) \cdot P(X_2=x_{23}) \cdot P(X_3=1) + P(r_{KII}=r_{KII3}|X_2=x_{24} \wedge X_3=1) \cdot P(X_2=x_{24}) \cdot P(X_3=1) +$
$P(r_{KII}=r_{KII3}|X_2=x_{25} \wedge X_3=1) \cdot P(X_2=x_{25}) \cdot P(X_3=1) + P(r_{KII}=r_{KII3}|X_2=x_{21} \wedge X_3=0) \cdot P(X_2=x_{21}) \cdot P(X_3=0) +$
$P(r_{KII}=r_{KII3}|X_2=x_{22} \wedge X_3=1) \cdot P(X_2=x_{22}) \cdot P(X_3=0) + P(r_{KII}=r_{KII3}|X_2=x_{23} \wedge X_3=1) \cdot P(X_2=x_{23}) \cdot P(X_3=0) +$
$P(r_{KII}=r_{KII3}|X_2=x_{24} \wedge X_3=1) \cdot P(X_2=x_{24}) \cdot P(X_3=0) + P(r_{KII}=r_{KII3}|X_2=x_{25} \wedge X_3=1) \cdot P(X_2=x_{25}) \cdot P(X_3=0).$

4 CONCLUSIONS

The developed model allows to determine the probability of the particular risk level. The result of modelling are the probability values for each risk level. Models can be modified for all the elements of the water supply system. Two models have been developed, for the first kind risk analysis and for the second kind risk analysis.

The risk assessment is based on the interpretation of the result (risk with the highest and the lowest probability of occurrence is given). For example, the result shows that for the first type of risk the highest probability is for a tolerable level and the lowest for the unacceptable level.

The developed model enables also determining the partial probabilities for events included in the defined Bayesian network.

REFERENCES

Apostolakis, G. & Kaplan, S. 1981. Pitfalls in risk calculations. *Reliability Engineering and System Safety* 2: 135–145.
Bishop, C.M. 2006. *Pattern Recognition and Machine Learning*. Springer: New York.
Bernardo, J.M. & Smith, A.F.M. 1993. *Bayesian theory*. Wiley: Chichester.
Boryczko, K. & Tchórzewska-Cieślak, B. 2013. Analysis and assessment of the risk of lack of water supply using the EPANET program, *Environmental Engineering IV*, Dudzińska M. R. Pawłowski L., Pawłowski A. (eds), Taylor & Francis Group, London: 63–68.
Boryczko, K. & Tchórzewska-Cieślak, B. 2012. Maps of risk in water distribution subsystem. *11th International Probabilistic Safety Assessment and Management Conference and the Annual European Safety and Reliability Conference, PSAM11-ESREL2012, June 25–29, 7, Helsinki, Finland*: 5832–5841.
Christodoulous, S., Charalambous, C. & Adamou, A. 2008. Rehabilitation and maintenance of water distribution network assets. *Water Science and Technology, Water Supply* 8(2): 231–237.

Hastak, H. & Baim, E. 2001. Risk factors affecting management and maintenance cost of urban infrastructure. *Journal of Infrastructure Systems* 7(2): 67–75.

Juraszka, B. & Braun, S. 2011. Practical Aspects of Operating a Water Treatment Plant on the Example of WTP Wierzchowo, *Rocznik Ochrona Srodowiska* 1: 973–988.

Kruszynski, W. & Dzienis, L. 2008. Selected Aspects of Modelling in Water Supply System on the Example of Lapy City. *Rocznik Ochrona Srodowiska* 10, 605–611.

Kwietniewski, M., Roman, M. & Kłoss-Trębaczkiewicz, H. 1993. *Niezawodność wodociągów i kanalizacji.* Wydawn. Arkady, Warszawa 1993.

Larose, D.T. 2006. *Odkrywanie wiedzy z danych.* Wprowadzenie do eksploracji danych. PWN: Warszawa.

Li, H., Apostolakis, G.E., Gifun J., Van, S.W., Leite, S. & Barber, D. 2009. Ranking the Risk from Multiple Hazards in a Small Community, *Risk Analysis* 3: 438–456.

Mazurkiewicz, J. & Walkowiak, T. 2004. Fuzzy economic analysis of simulated discrete transport system. *7th International Conference on Artificial Intelligence and Soft Computting- ICAISC, Poland*, Zakopane. Lecture Notes in Artificial Intelligence 3070: 1161–1167.

Morzy, M. 2007. Eksploracja danych. *Nauka* 3: 83–104.

Pietrucha-Urbanik K. & Tchórzewska-Cieślak B. 2014. Water Supply System operation regarding consumer safety using Kohonen neural network; in: Safety, Reliability and Risk Analysis: Beyond the Horizon – Steenbergen et al. (Eds), Taylor & Francis Group, London: 1115–1120.

Pollard, S.J.T., Strutt, J.E., Macgillivray, B.H., Hamilton, P.D. & Hrudey, S.E. 2004. Risk analysis and management in the water utility sector – a review of drivers, tools and techniques. *Process Safety and Environmental Protection* 82(6): 1–10.

Rak, J. 2009. Selected problems of water supply safety. *Environmental Protection Engineering* 35: 29–35.

Rak, J.R. & Pietrucha, K. 2008. Risk in drinking water control. Przemysł Chemiczny 87(5): 554–556.

Ritter, G. & Gallegos, T. 2002. Bayesian object identification: variants, *Journal of Multivariate Analysis* 81: 301–334.

Studzinski, A. & Pietrucha-Urbanik, K. 2012. Risk Indicators of Water Network Operation. *Chemical Engineering Transactions* 26: 189–194. DOI: 10.3303/CET1226032.

Tchorzewska-Cieslak, B. 2013. Bayesian Model of Urban Water Safety Management. Proceedings of the 13th International Conference of Environmental Science and Technology Athens, Greece, 5–7 September 2013.

Tchórzewska-Cieślak, B. & Włoch, A. 2006. Method for risk assessment in water supply systems. *4th International Probabilistic Symposium. October 12–13 2006 BAM Berlin*: 279–288.

Tchórzewska-Cieślak, B. 2011. *Methods for analysis and assessment of risk in water distribution subsystem* (in polish). Oficyna Wydawnicza Politechniki Rzeszowskiej: Rzeszów.

Thompson, W.E. & Springer, M.D. 1972. Bayes analysis of availability for a system consisting of several independent subsystems. *IEEE Transactions on Reliability* 21(4): 212–218.

Valis, D., Vintr, Z. & Koucky, M. 2010. Contribution to highly reliable items' reliability assessment. *Proc. of the European Safety and Reliability Conference ESREL, Prague, Czech Republic. Reliability, Risk and Safety: Theory and Applications.* Taylor & Francis 1–3: 1321–1326.

Walkowiak, T. & Mazurkiewicz, J. 2009. Analysis of Critical Situations in Discrete Transport Systems. *Proc. of the International Conference on Dependability of Computer Systems.* DOI: 10.1109/DepCoS-RELCOMEX.2009.39. Poland, Brunow: 364–371.

Zitrou, A., Bedford, T. & Walls, L. 2010. Bayes geometric scaling model for common cause failure rates. *Reliability Engineering & System Safety* 95(2): 70–76. DOI: 10.1016/j.ress.2009.08.002.

Zhang, T.L. & Horigome, M. 2001. Availability and reliability of system with dependent components and time-varying failure and repair rates. *IEEE Transactions on Reliability* 50(2): 151–158. DOI: 10.1109/24.963122.

Progress in Environmental Engineering – Tomaszek & Koszelnik (eds)
© *2015 Taylor & Francis Group, London, ISBN: 978-1-138-02799-2*

Differentiation of selected components in bottom sediments of Poland's Solina-Myczkowce complex of dam reservoirs

L. Bartoszek & J.A. Tomaszek
Department of Chemistry and Environmental Engineering, Rzeszów University of Technology, Rzeszów, Poland

J.B. Lechowicz
Department of Industrial and Materials Chemistry, Rzeszów University of Technology, Rzeszów, Poland

ABSTRACT: Statistical analysis was applied to compare mean contents of total phosphorus, iron, manganese, aluminium, calcium and organic matter, as well as pH, in bottom sediments of two Polish reservoirs (Solina and Myczkowce). It proved possible to observe natural spatial differentiation in the chemical compositions of the bottom sediments in different parts of the same reservoir, and also between reservoirs. In the case of such a large body of water as the Solina Reservoir (and despite relatively limited differences in means of land management and utilisation), the influence of the drainage basin on bottom sediments in the zone of the reservoir influenced by river flow was seen to be rather marked.

1 INTRODUCTION

A dominant process of the cycling of substances in dam reservoirs is sedimentation. The greater part of the material brought in by river waters is in suspension, hence a slackening of the current at the point of influx into a reservoir results in a decline in the capacity of the water to carry sediment, the effect being the formation of sediment from the allochthonous material arriving from the river basin, along with the autochthonous matter generated in the course of primary or secondary production (Wiśniewski 1995, Kentzer 2001, Borówka 2007). Heavier particles are dropped in the upper part of a reservoir, while successively lighter and finer ones are carried further towards the dam. The intensity of sedimentation is thus mainly related to the type of suspension (mineral v. organic, large v. small, heavy v. light), the time for which water is retained in the reservoir, and chemical conditions (favouring the formation and precipitating out of weakly-soluble phosphorus, iron, calcium, aluminium and manganese compounds) (House & Denison 2000, Bajkiewicz-Grabowska 2002, Håkanson & Jansson 2002, Lehtoranta & Pitkänen 2003). The result of the process is the laying-down of substances in bottom sediments in quantities that far exceed those in the water column (Wiśniewski 1995).

In the upper parts of dam reservoirs, deposits are like those in rivers (with typical sandy river sediments prevailing). In contrast, in the central and lower (near-dam) areas, even where the throughput is considerable, there are rather muddy deposits similar in nature to the gyttja present in lakes. Since they vary considerably in thickness, bottom sediments may contain quite disparate amounts of elements. It is a usual circumstance for reservoirs that, the deeper the water, the greater the degree to which sediments comprise fine particles and are present in greater thicknesses, also mostly containing more phosphorus and organic matter (Borówka 2007). The thickness of the surface layer of sediment most actively exerting an impact on the near-bottom water is estimated at several centimetres, the key determining feature being the degree of hydration of the deposit (Żbikowski 2004).

In large dam reservoirs, the spatial differences in environmental and biotic conditions may be very considerable (Watts 2000). The chemical composition of the bottom sediments of a body of water depend to a significant degree on characteristics of its basin, as well on the means of utilization and management (Müller et al. 1998, Ankers et al. 2003, Mielnik 2005). The level of pollution of the sediments may be considered an indicator of how the ecosystem is loaded with different substances, of anthropogenic origin in particular (Borówka 2007, Anderson & Pacheco 2011).

The work described here had as its aim an analysis of spatial differentiation to contents of selected components in the bottom sediments of the Solina-Myczkowce complex of dam reservoirs, which includes a main reservoir (Solina) and a top-up reservoir (Myczkowce). Selected elements (especially iron, aluminium, manganese and calcium) control the flow of phosphorus in water reservoirs under natural conditions and precipitating out in form of weakly-soluble compounds. In the case of such a large object, which is the Solina Reservoir, despite minor differences in the way of development and land use, catchment has a significant influence on the composition of sediments especially in the zone of the river influence.

2 MATERIALS AND METHODS

2.1 Study sites

The Solina Reservoir is the largest dam reservoir in Poland in volume terms, and also the deepest. It joins the Myczkowce Reservoir within the framework of the hydroelectric power company known as *Zespół Elektrowni Wodnych Solina-Myczkowce S.A.* Myczkowce is the top-up reservoir for the operations of a pumped-storage power station, ensuring that Solina and Myczkowce are in fact two very different bodies of water in terms of their morphometric parameters (Table 1). The main supply of the Myczkowce Reservoir originate from the San River (over 90%), which arrive via the hypolimnion water of the Solina Reservoir (Koszelnik 2009a). The basin of the Solina-Myczkowce Reservoirs is mainly forest land with only limited settlement or agricultural use. Tourist and settlement infrastructure is mainly located in the near-confluence areas of tributaries and in the basin areas immediately around the bodies of water.

2.2 Sediment sampling and analyses

Samples of bottom sediment were collected from four sites in the Solina Reservoir, known as: 1. Centralny (the "central" site), and 2. Zapora, 3. Brama and 4. Skałki, as named after localities (Fig. 1). The sites are characterised by depths of ca. 45, 55, 14 and 15 m respectively. In addition, there were two sampling sites at the Myczkowce Reservoir, i.e. 5. Myczk. Zapora and 6. Myczk. Zabrodzie at depths of around 11 and 3 m respectively. The sampling was done once or twice a month in the May–November period of 2005, as well as once a month in the April–November period of 2006, excluding May (16 series in all, except sites 2 and 6 with 15 series).

Table 1. Morphometric characteristics of the Solina-Myczkowce complex of dam reservoirs (Koszelnik 2009a).

Feature	Solina Reservoir	Myczkowce Reservoir
Area [ha]	2200	200
Maximum volume [M m^3]	502	10
Average (max.) depth [m]	22(60)	5(15)
Catchment area [km^2]	1174.5	1248
Water retention time [d]	155–273	2–6

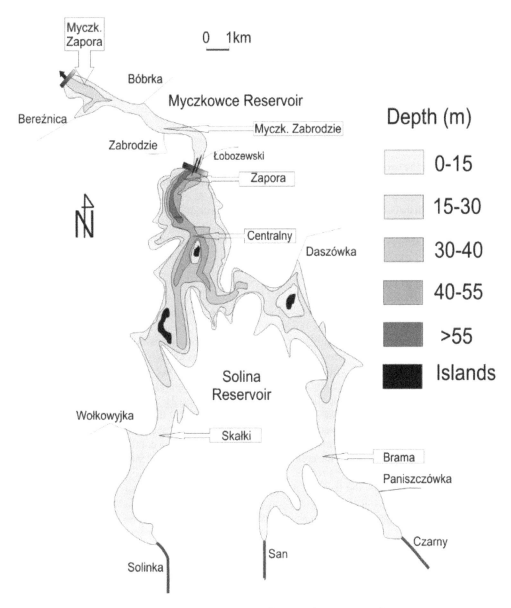

Figure 1.   Distribution of measurement points in the Solina-Myczkowce Reservoirs.

The 0–5 cm superficial layer was taken for analysis, averages being calculated for three sediment cores sampled with a gravity corer. Interstitial water was separated out from the samples prepared in this way using centrifugation at 4000 revolutions per min. The sediment obtained was then air-dried at room temperature, as well as 60°C, before being broken up fully and sieved. The fraction of grain size below 0.9 mm was retained for study in sealed PE bags at a temperature of 4°C and in the dark. The sediments were mineralised thereafter using concentrated HNO_3 (microwave digestion method at high pressure 2–4.5 MPa – UniClever II, Plazmatronika).

The main methods used in analysing the variables under study were colorimetric: PN-EN 1189:2000 (for phosphorus), PN-ISO 6332:2001 (iron), DIN ISO 10566E30 (aluminium) and DIN

13

38406E2 (manganese). Colorimetric determinations were carried out using an Aquamate spectrophotometer (Thermo Spectronic, United Kingdom). The contents of calcium in the mineralised samples were determined by means of AAS (Perkin Elmer, AAnalyst 300), organic matter (OM) in sediments by oxidation at 550°C for 4 h, and sediment pH (pH_{KCl}) potentiometrically in a 1 mol dm^{-3} colloidal suspension with KCl. Each sample was subject to three replicate sets of determination, the ultimate result being the mean deriving from values not differing from one another by more than 10% from the lower one.

2.3 Statistical analyses

Mean values for the two groups were compared using the Student t test, the Cochran–Cox test (t test with separate analysis of variance), and the non-parametric Kolmogorov–Smirnov test. Analysis of differences between mean values in several groups made use of ANOVA (with the Shapiro-Wilk test for normal distributions, Levene's test for equality of variances, and the Fisher–Snedecor test, as well as the parametric Scheffé test and the non-parametric Kruskal–Wallis test). In each case the adopted significance level was 0.05 (Stanisz 1998, Bartoszek 2008).

3 RESULTS AND DISCUSSION

Statistical analysis of mean concentrations of selected elements pointed to very significant spatial differences in contents of most of them, this applying between bodies of water, between zones within the Solina Reservoir, and between different research sites. Across the whole research period, mean contents of total P in deposits were slightly higher in sediments collected from the lacustrine zone than in those of the river flows within the Solina Reservoir (Bartoszek & Tomaszek 2011). Similar trends were also to be noted as regards the concentrations of iron, aluminium and manganese in deposits, while the reverse trend applied to calcium content (Fig. 2). Statistical tests (the Student t test, Cochran–Cox and Kolmogorov–Smirnov tests) confirmed that the sediments of the shallower and deeper parts of the Solina Reservoir did differ significantly (test probability values $p < 0.05$) when it came to phosphorus, aluminium, iron, manganese and calcium contents, as well as sediment pH (pH_{KCl}). In turn, in the case of the content of organic matter, the Student t test did not reveal any statistically significant differences related to the depth in the Solina Reservoir at which deposits were sampled. This despite the fact that sediments collected from the greatest depths are usually found to have the greatest accumulations of organic matter (Trojanowski & Antonowicz 2005).

ANOVA for mean concentrations in sediments from the different sites was able to confirm that deposits differ significantly as regards the content of determined components. However, in relation to given components, similarities and differences between sites did not seem to follow a regular pattern, and a distinction between one reservoir and the other can often not be drawn. The lowest mean content obtained for total P was the 0.689 mg g^{-1} of d.w. observed in sediments at the Skałki site, which is within the zone of river influence. The value in question was significantly different from those obtained for the remaining deposits studied within the Solina Reservoir, as well as those taken from the Myczk. Zapora site. It was in turn most similar to the value noted for phosphorus in the Myczk. Zabrodzie sediments (i.e. 0.754 mg g^{-1} of d.w.) (Fig. 2). In turn, in the deposits from the Brama site (also under the influence of river inflows), the total P content was higher (at 0.857 mg g^{-1} of d.w.), and hence close to those noted within the reservoirs' lacustrine zones. The mean concentrations of total P in sediments from the Centralny, Zapora and Myczk. Zapora sites (i.e. 0.912; 0.931 and 0.869 mg g^{-1} of d.w. respectively) did not differ significantly (Table 2). The higher phosphorus content in the sediments collected from reservoir lacustrine zones might have been the effect of enhanced sedimentation of autochthonous material containing the element (Wiśniewski 1995, Moosmann et al. 2006). Silty sediments of lacustrine zone, due to the smaller particle size also have a greater specific surface area and thus a greater capacity for adsorption of dissolved constituents in water. The Myczkowce Reservoir is found to be characterised by à significantly different total phosphorus content in its deposits. There was an analogous situation

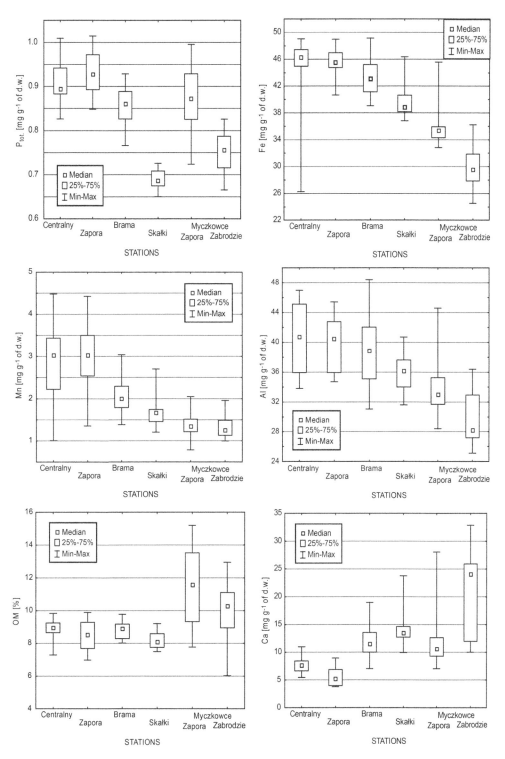

Figure 2.   Statistical distribution to contents of total phosphorus ($P_{tot.}$), iron, manganese, aluminium, calcium [mg g^{-1} of d.w.], and organic matter (OM) [%] in the bottom sediments of the Solina–Myczkowce dam reservoirs.

in the top-up reservoir as in the main one, in that the mean value for total P content was higher in sediments taken from the deeper part of the reservoir than in those from Myczk. Zabrodzie. A similar trend could also be observed in the case of the iron and aluminium contents in deposits, while the reverse situation applied to contents of calcium (Fig. 2).

Bottom sediments from the different sites in the two reservoirs were shown to differ significantly in their contents of iron and manganese. The highest contents of the two elements in deposits of the Solina Reservoir were those at the Zapora site, the lowest those at Skałki (Fig. 2). The non-parametric Kruskal-Wallis test pointed to significant differences in contents of the above-mentioned parameters between the sediments at the two sites, while not revealing any statistical differences between the remaining deposits in the Solina Reservoir, or between the deposits in the Myczkowce Reservoir (Table 2). Where the iron content was concerned, sediments from the Myczk. Zapora site were significantly similar to those from the Myczk. Zabrodzie and Skałki sites, nevertheless manifesting statistically significant differences when compared with the Solina Reservoir's remaining deposits. From the point of view of the content of manganese, the sediments of the top-up reservoir were similar to those of the Skałki site, while differing significantly from the deposits at the Centralny, Zapora and Brama sites in the main reservoir. In the cases of the aluminium content of sediments, no significant differences were noted between deposits within given reservoirs, the only ones noted being between the bodies of water (Table 2). Sediments from the Myczk. Zapora site manifested similarity from the point of view of aluminium content to those from the shallower parts of the Solina Reservoir (Skałki and Brama sites).

Organic matter content did not display statistically significant differences within the deposits from the two reservoirs (Table 2). However, there were significant differences between the lowest and highest contents as noted in the sediments from the Skałki site on the one hand and the sediments from the Myczkowce Reservoir on the other (Fig. 2). Determined by Koszelnik (2009b), accumulation of organic matter of autochthonous origin in the Solina Reservoir lacustrine zone was about $987 \, t \, yr^{-1}$. Organic matter derived from the production inside the reservoirs was about 70% of the total accumulated matter. The generally low contents for OM may be linked with relatively impoverished trophic status, as well as intensive mineralisation in the well-oxygenated conditions present in the near-bottom zone (Czarnecka et al. 2005, Moosmann et al. 2006). Moreover, the degree of mineralisation of sediments may increase in deep reservoirs where the sedimentation time is longer (Kentzer 2001).

The attention is drawn to the statistically-significant differences revealed by ANOVA in the case of the calcium content of the deposits. A division into two groups of sediment could be noted, i.e. (1) sediments from the shallower parts of the Solina Reservoir (Brama and Skałki sites), as well as the Myczkowce Reservoir, and (2) sediments from the deeper parts of the Solina Reservoir (Centralny and Zapora sites) (Table 3). In terms of the sediment pH a very clear division into three groups of deposits could be noted, i.e. deposits: (1) from the Centralny and Zapora sites, which is to say the deeper parts of the Solina Reservoir, and hence its lacustrine zone, (2) from the Brama and Myczk. Zapora sites, and (3) from the Skałki and Myczk. Zabrodzie sites.

When an analysis of the mean values for the two variables was carried out, it was noticeable that the lowest contents of calcium and a lower (slightly acid) reaction ($pH_{KCl} < 7$) was characteristic of the deposits of the lacustrine zone in the Solina Reservoir (Figs. 2–3). A similar content of calcium and almost the same neutral reaction ($pH_{KCl} \sim 7$) was in turn characteristic of the sediments from the Brama and Myczk. Zapora sites. Finally, the highest values for calcium content combined with a higher pH value over 7 applied in the case of the deposits from the Skałki and Myczk. Zabrodzie sites. In the cooler hypolimnion (the deeper parts of the reservoirs), the higher concentrations of CO_2 present may give rise to a situation in which part of the sedimenting calcium carbonate (IV) may be dissolved, with the result that only some is deposited as sediment. In the course of an intensive process of organic matter breakdown, both organic acids and CO_2 are liberated, the result being a lowering of the sediment pH and interstitial water alike (Golterman 2005). This lowering of pH may also influence the dissolving of calcium compounds deposited in the lacustrine-zone sediments (Borgnino et al. 2006). Higher contents of calcium compounds may also counteract the lowering of sediment pH in the shallower parts of the Solina and Myczkowce Reservoirs. Calcium

Table 2. Matrices of estimate of the statistically significant differences in the mean contents of total phosphorus ($P_{tot.}$), iron, manganese, aluminium [$mg \cdot g^{-1}$ of d.w.], and organic matter (OM) [%] in the bottom sediments. Test probability values $p < 0.05$ – mean values are different. H-value of Kruskal-Wallis test.

	Independent variable – Station					
	Centralny	Zapora	Brama	Skałki	Myczk. Zapora	Myczk. Zabrodzie
Dependent variable P_{tot}	Kruskal–Wallis test: H (5, n = 94) = 65.55 p < 0.05					
Centralny		1.0000	1.0000	**0.0000**	1.0000	**0.0002**
Zapora	1.0000		0.2922	**0.0000**	0.9474	**0.0000**
Brama	1.0000	0.2922		**0.0003**	1.0000	0.1032
Skałki	**0.0000**	**0.0000**	**0.0003**		**0.0000**	1.0000
Myczk. Zapora	1.0000	0.9474	1.0000	**0.0000**		**0.0220**
Myczk. Zabrodzie	**0.0002**	**0.0000**	0.1032	1.0000	**0.0220**	
Dependent variable Fe	Kruskal–Wallis test: H (5, n = 94) = 66.00 p < 0.05					
Centralny		1.0000	1.0000	0.0653	**0.0000**	**0.0000**
Zapora	1.0000		1.0000	**0.0478**	**0.0000**	**0.0000**
Brama	1.0000	1.0000		1.0000	**0.0068**	**0.0000**
Skałki	0.0653	**0.0478**	1.0000		1.0000	**0.0100**
Myczk. Zapora	**0.0000**	**0.0000**	**0.0068**	1.0000		1.0000
Myczk. Zabrodzie	**0.0000**	**0.0000**	**0.0000**	0.0100	1.0000	
Dependent variable Mn	Kruskal–Wallis test: H (5, n = 94) = 49.42 p < 0.05					
Centralny		1.0000	1.0000	0.0900	**0.0003**	**0.0002**
Zapora	1.0000		0.9224	**0.0036**	**0.0000**	**0.0000**
Brama	1.0000	0.9224		1.0000	**0.0113**	**0.0087**
Skałki	0.0900	**0.0036**	1.0000		1.0000	1.0000
Myczk. Zapora	**0.0003**	**0.0000**	**0.0113**	1.0000		1.0000
Myczk. Zabrodzie	**0.0002**	**0.0000**	**0.0087**	1.0000	1.0000	
Dependent variable Al	Kruskal–Wallis test: H (5, n = 94) = 44.38 p < 0.05					
Centralny		1.0000	1.0000	0.2431	**0.0042**	**0.0000**
Zapora	1.0000		1.0000	0.4871	**0.0121**	**0.0000**
Brama	1.0000	1.0000		1.0000	0.0974	**0.0001**
Skałki	0.2431	0.4871	1.0000		1.0000	**0.0385**
Myczk. Zapora	**0.0042**	**0.0121**	0.0974	1.0000		1.0000
Myczk. Zabrodzie	**0.0000**	**0.0000**	**0.0001**	**0.0385**	1.0000	
Dependent variable OM	Kruskal–Wallis test: H (5, n = 94) = 29.55 p < 0.05					
Centralny		1.0000	1.0000	0.4102	0.1431	1.0000
Zapora	1.0000		1.0000	1.0000	**0.0041**	0.1918
Brama	1.0000	1.0000		0.7552	0.0673	1.0000
Skałki	0.4102	1.0000	0.7552		**0.0000**	**0.0045**
Myczk. Zapora	0.1431	**0.0041**	0.0673	**0.0000**		1.0000
Myczk. Zabrodzie	1.0000	0.1918	1.0000	**0.0045**	1.0000	

provides buffering that is capable of obstructing any more major changes in pH. A higher calcium content in sediments of the zone under the influence of a river may also be associated with the sedimentation here of heavier allochthonous material that is rich in calcium and has arisen thanks to the rinsing and leaching of agricultural soils.

Table 3. Matrices of estimate of the statistically significant differences in the mean contents of calcium [mg·g⁻¹ of d.w.] in the bottom sediments and pH of sediment. Test probability values p < 0.05 – mean values are different. H-value of Kruskal-Wallis test.

	Independent variable – Station					
	Centralny	Zapora	Brama	Skałki	Myczk. Zapora	Myczk. Zabrodzie
Dependent variable Ca	Kruskal–Wallis test: H (5, n = 94) = 63.88 p < 0.05					
Centralny		1.0000	**0.0396**	**0.0002**	0.0554	**0.0000**
Zapora	1.0000		**0.0003**	**0.0000**	**0.0005**	**0.0000**
Brama	**0.0396**	**0.0003**		1.0000	1.0000	0.2585
Skałki	**0.0002**	**0.0000**	1.0000		1.0000	1.0000
Myczk. Zapora	0.0554	**0.0005**	1.0000	1.0000		0.1951
Myczk. Zabrodzie	**0.0000**	**0.0000**	0.2585	1.0000	0.1951	
Dependent variable pH	Scheffé test: p < 0.05					
Centralny		0.9232	**0.0000**	**0.0000**	**0.0001**	**0.0000**
Zapora	0.9232		**0.0000**	**0.0000**	**0.0000**	**0.0000**
Brama	**0.0000**	**0.0000**		**0.0274**	0.9970	**0.0195**
Skałki	**0.0000**	**0.0000**	**0.0274**		**0.0056**	1.0000
Myczk. Zapora	**0.0001**	**0.0000**	0.9970	**0.0056**		**0.0038**
Myczk. Zabrodzie	**0.0000**	**0.0000**	**0.0195**	1.0000	**0.0038**	

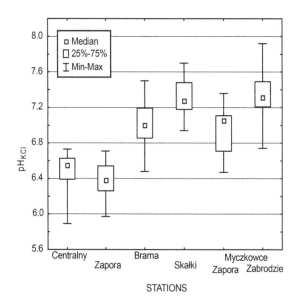

Figure 3. Statistical distribution of sediment pH (pH$_{KCl}$) in the Solina–Myczkowce dam reservoirs.

The sediments of the Solina Reservoir differ from those of the Myczkowce Reservoir in their higher mean contents of total phosphorus, iron, aluminium and manganese, as well as lower OM and calcium contents and a more acid reaction (Table 4). Statistical analysis (based on the Student t test, Cochran–Cox test and Kolmogorov–Smirnov test) confirmed the occurrence of statistically significant differences between the means in the two studied groups (at p < 0.05), between the contents noted for total P, aluminium, iron, manganese, calcium, organic matter and

Table 4. The contents of total phosphorus (Ptot.), iron, aluminium, manganese and calcium [mg · g^{-1} of d.w.], and organic matter (OM) [%] in the bottom sediments of the Solina – Myczkowce Reservoirs, as well as sediment pH, and mean contents of certain components in the bottom sediments of the Solina–Myczkowce Reservoirs in the years 1988–1989 (Tomaszek & Czerwieniec 1996).

		P_{tot} [mg g^{-1} of d.w.]	Fe	Al	Mn	Ca	OM [%]	pH [pH]
Centralny n = 16	mean	0.912	44.6	40.5	2.82	7.67	8.86	6.48
Zapora n = 15	mean	0.931	45.6	39.9	3.05	5.56	8.48	6.37
Brama n = 16	mean	0.857	43.3	38.9	2.10	11.96	8.81	7.01
Skałki n = 16	mean	0.689	39.8	35.8	1.66	14.25	8.19	7.32
Myczk. Zapora n = 16	mean	0.869	35.9	34.1	1.37	12.90	11.52	6.96
Myczk. Zabrodzie n = 15	mean	0.754	29.9	29.7	1.35	20.8	9.74	7.34
Solina n = 63 2005–2006	mean	0.846	43.3	38.8	2.40	9.93	8.59	6.80
	median	0.873	44.7	38.3	2.17	9.00	8.63	6.71
	SD	0.11	4.1	4.4	0.9	4.3	0.7	0.4
Myczkowce n = 31 2005–2006	mean	0.813	33.0	32.0	1.36	16.7	10.7	7.14
	median	0.797	34.1	32.0	1.34	12.4	10.4	7.14
	SD	0.09	4.3	4.4	0.3	7.8	2.4	0.3
Solina 1988–1989	mean	0.56	38.8	–	–	11.7	8.79	–
Myczkowce 1988–1989	mean	0.55	32.5	–	–	12.7	9.1	–

pH in the sediments of the main reservoir and the top-up reservoir. Waters of the hypolimnion of the Solina Reservoir may be subject to the sedimentation of organic matter and the products of its decomposition, leading to an enrichment of the waters of the top-up reservoir in this substance.

In the Myczkowce Reservoir, due to very short retention time of water, the increase of deposits, including the organic matter was probably caused mainly by the growth and death of macrophytes (Koszelnik 2009b). The mean content of iron in the deposits of the reservoirs studied (Table 4) was more than four times as high as the mean content (7.5 mg g^{-1} of d.w.) given for bottom sediments of Polish reservoirs, and was thus close to the mean content (of 35.9 mg g^{-1} of d.w.) cited for the bottom sediments of reservoirs in the world as a whole (Wiechuła 2004). Furthermore, the mean contents given for manganese in Polish reservoirs (0.26 mg g^{-1} of d.w.) and those around the world as a whole (0.42 mg g^{-1} of d.w.) (Wiechuła 2004) were many times exceeded (especially in the deposits of the upper reservoir). Certain quantities of iron may derive from the decomposition of organic matter, while the primary sources of the element – as well as of manganese – are most often weathering minerals from the magmatic rocks present in the basin (Czamara & Czamara 2008). The high iron content in the sediments of the above reservoirs was already apparent when research was carried out in the years 1988–1989 (Table 4) (Tomaszek & Czerwieniec 1996). Equally, the ca. 17-year period brought increases in the content of total phosphorus in the sediments of each reservoir. The content of organic matter was also higher than before in the deposits of the Myczkowce Reservoir. Concentrations of iron rose slightly in the sediments of the Solina Reservoir, while calcium currently shows a greater range of values between deposits in the two bodies of water analysed.

4 CONCLUSIONS

The bottom sediments of the Solina–Myczkowce Reservoir complex are mainly mineral in nature, and are rich in iron, aluminium and manganese, as well as relatively poor in phosphorus, calcium and organic matter. Beyond that, it was possible to observe natural spatial differentiation to the chemical composition of sediments in different parts of the same reservoir, as well as between reservoirs.

In the case of a large reservoir of the kind the Solina Reservoir represents (and notwithstanding the relatively limited differences in means of managing and utilising land), the influence of the river basin on bottom sediments in the river inflow zone turns out to be very visible. The cause of the observed differences in the chemical composition of sediments collected in the river inflow zone of the main reservoir is the greater degree of land management, as well as the agricultural use made of the San River and Czarny Stream basins, as opposed to the Solinka River.

The sediments of the Myczkowce Reservoir differ from those of the Solina Reservoir in that they have higher mean contents of organic matter and calcium, as well as a higher pH, and then lower contents of total P, iron, aluminium and manganese. The composition of bottom sediments in the top-up reservoir is mainly shaped from components deriving from water coming in from the main reservoir located above. The marked intensity of the process of water exchange at Myczkowce in turn ensures that the retention of mineral substances may mainly take place in the zone of direct contact between deposits and near-bottom water.

The sediments in the reservoirs studied are in large measure of allochthonous origin, as is indicated by their high content of mineral matter. The generally low contents for OM and calcium may be linked with relatively impoverished trophic status especially of the upper reservoir.

Acknowledgements Research part-financed by the Ministry of Science and Higher Education within the framework of the PO4G 084 27 and N523 009 32/0288 projects. We thank the members of the projects for their cooperation.

REFERENCES

Anderson, M.A. & Pacheco, P. 2011. Characterization of bottom sediments in lakes using hydroacoustic methods and comparison with laboratory measurements. *Water Res* 45: 4399–4408.

Ankers, C., Walling, D.E., Smith, R.P. 2003. The influence of catchment characteristics on suspended sediment properties. *Hydrobiologia* 494: 159–167.

Bajkiewicz-Grabowska, E. 2002. *Obieg materii w systemach rzeczno-jeziornych*. Warszawa: Uniwersytet Warszawski. Wydział Geografii i Studiów Regionalnych.

Bartoszek, L. 2008. Badania retencji związków fosforu w osadach dennych na przykładzie zbiorników zaporowych Solina – Myczkowce. *Dysertacja doktorska*. Wydział Inżynierii Srodowiska Politechniki Lubelskiej. Rzeszów.

Bartoszek, L. & Tomaszek, J.A. 2011. Analysis of the spatial distribution of phosphorus fractions in the bottom sediments of the Solina – Myczkowce dam reservoir complex. *Environ Prot Eng* 37(3): 5–15.

Borgnino, L., Avena, M., De Pauli, C. 2006. Surface properties of sediments from two Argentinean reservoirs and the rate of phosphate release. *Water Res* 40: 2659–2666.

Borówka, R.K. 2007. Geochemiczne badania osadów jeziornych strefy umiarkowanej. *Studia Limnologica et Telmatologica* 1(1): 33–42.

Czamara, A. & Czamara, W. 2008. Metale ciężkie w systemie ekologicznym zbiornika Mściwojów. *Infrastruktura i Ekologia Terenów Wiejskich* 9: 283–296.

Czarnecka, K., Tylmann, W., Woźniak, P.P. 2005. Recent sediments of Lake Druzno (southern basin). *Limnological Review* 5: 45–51.

Golterman, H.L. 2005. *The Chemistry of Phosphate and Nitrogen Compounds in Sediments*. Kluwer Academic Publisher.

Håkanson, L. & Jansson, M. 2002. *Principles of Lake Sedimentology*. New Jersey: The Blackburn Press.

House, W.A. & Denison, F.H. 2000. Factors influencing the measurement of equilibrium phosphate concentrations in river sediments. *Water Res* 34(4): 1187–1200.

Kentzer, A. 2001. *Fosfor i jego biologicznie dostępne frakcje w osadach jezior różnej trofii*. Toruń: Wydawnictwo UMK.

Koszelnik, P. 2009a. Isotopic effects of suspended organic matter fluxes in the Solina reservoir (SE Poland). *Environ Prot Eng* 35(4): 13–19.

Koszelnik, P. 2009b. *Źródła i dystrybucja pierwiastków biogennych na przykładzie zespołu zbiorników zaporowych Solina-Myczkowce*. Rzeszów: Oficyna Wydawnicza Politechniki Rzeszowskiej.

Lehtoranta, J. & Pitkänen, H. 2003. Binding of phosphate in sediment accumulation areas of the eastern Gulf of Finland, Baltic Sea. *Hydrobiologia* 492: 55–67.

Mielnik, L. 2005. Wpływ użytkowania zlewni na właściwości fizykochemiczne osadów dennych jezior. *Inżynieria Rolnicza* 4(64): 31–36.

Moosmann, L., Gächter, R., Müller, B., Wüest, A. 2006. Is phosphorus retention in autochthonous lake sediments controlled by oxygen or phosphorus?. *Limnol Oceanogr* 51(1): 763–771.

Müller, B., Lotter, A.F., Sturm, M., Ammann, A. 1998. Influence of catchment quality and altitude on the water and sediment composition of 68 small lakes in Central Europe. *Aquat Sci* 60: 316–337.

Stanisz, A. 1998. *Przystępny kurs statystyki z wykorzystaniem programu STATISTICA na przykładach z medycyny. Tom I. Statystyki podstawowe.* Kraków: StatSoft Poland.

Tomaszek, J.A. & Czerwieniec, E. 1996. The chemistry of bottom sediments of the Solina reservoir. In W.M. Harasimiuk & J. Wojtanowicz (eds), *Regional Ecological Problems, Part 1. Scientific Works of the IIAREP*: 181–190.

Trojanowski, J. & Antonowicz, J. 2005. Właściwości chemiczne osadów dennych jeziora Dołgie Wielkie. *Słupskie Prace Biologiczne* 2: 123–133.

Watts, C.J. 2000. Seasonal phosphorus release from exposed, re-inundated littoral sediments of two Australian reservoirs. *Hydrobiologia* 431: 27–39.

Wiechuła, D. 2004. *Ekotoksykologia osadów dennych zbiorników o różnej charakterystyce limnologicznej.* Katowice: Wydawnictwo Śląska Akademia Medyczna.

Wiśniewski, R.J. 1995. Rola zasilania wewnętrznego w eutrofizacji zbiorników zaporowych. In M. Zalewski (ed), *Procesy biologiczne w ochronie i rekultywacji nizinnych zbiorników zaporowych*: 61–70. Łódź: Biblioteka Monitoringu Środowiska.

Żbikowski, J. 2004. Rola fauny dennej w wymianie substancji w interfazie: osad – woda. In R. Wiśniewski & J. Jankowski (eds), *Ochrona i Rekultywacja Jezior*: 253–260. Toruń: PZITS.

Progress in Environmental Engineering – Tomaszek & Koszelnik (eds)
© 2015 Taylor & Francis Group, London, ISBN: 978-1-138-02799-2

The role of wetlands in the removal of heavy metals from the leachate (on the example of the Lipinka River catchment, southern Poland)

T. Molenda

Faculty of Earth Sciences, University of Silesia, Sosnowiec, Poland

ABSTRACT: Industrial waste landfills are a significant source of contamination of the surface and groundwater in their storage area. Contamination of the hydrosphere is mainly a consequence of the formation of a highly mineralised leachate. This paper presents the impact of an industrial waste landfill on the contamination of the Lipinka River. This river receives the leachate from a landfill of metallurgical slag resulting from the processing (smelting) of zinc-lead ores (Zn-Pb). The leachate is characterised by a high electrolytic conductivity ($K_{25} - 7644\,\mu S/cm$) and a high concentration of heavy metals (Zn – 8.1 mg/L; Cd – 0.062 mg/L; Pb – 0.015 mg/L; Cu – 0.015 mg/L). The supply of the leachate into the river contaminates its waters with heavy metals. It has been observed that the movement of the contaminated water through the reservoir covered with willow moss (*Fontinalis antipyretica*) has a significant impact on the removal of zinc and cadmium from the water. There has been a significant decrease in the concentration of zinc and cadmium as well as a reduction of the load of metals. The removal of zinc in the reservoir ranged from 54 to 81%, while the removal of cadmium ranged from 64 to 90%. The research has indicated that willow moss can be used in the treatment of hydrophytic industrial wastewater.

1   INTRODUCTION

Industrial waste landfills are a significant source of contamination of both surface and groundwater. Contamination of the hydrosphere is mainly a result of the formation of a highly mineralised leachate. In addition to the commonly occurring ions, such as calcium and magnesium, the leachate may contain high concentrations of other, often toxic elements (Twardowska 1981, Szczepańska & Twardowska 1999, Szczepańska & Twardowska 2004, Stefaniak & Twardowska 2006, Molenda 2006, Molenda & Chmura 2007, Molenda & Chmura 2012).

In the Upper Silesian Coal Basin (southern Poland) (Fig. 1), the main sources of contamination of the hydrosphere are the coal mining landfills which hold 632 718 400 000 tons of waste. There are also a number of landfills connected with the extraction and processing of sulphide ores in the region (Cabała 2005, Cabała et al. 2007, Cabała 2009). The impact of these landfills on the aquatic environment is described, among others, in Jankowski et al. (2006), Molenda (2006), Jonczy (2006).

This paper describes the impact of the metallurgical slag landfills resulting from the smelting of the zinc-lead ores on the contamination of the water in the Lipinka River. The aim of the study was to demonstrate what impact an anthropogenic wetland exerts on the removal of heavy metals from the water.

2   LOCATION OF THE RESEARCH AREA

The catchment of the Lipinka River is located in southern Poland in the Upper Silesian Coal Basin (Fig. 1). The geometric centre of the basin lies on the coordinates of $50°18'53.97''N$ and

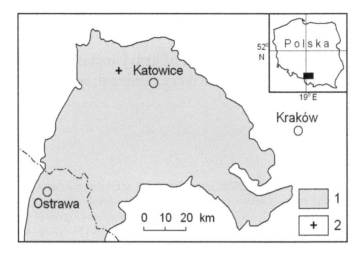

Figure 1.   Location of the research area. Key: 1 – the area of the Upper Silesian Coal Basin, 2 – the position of the catchment of the Lipinka River. The landfill along with the height of the edge.

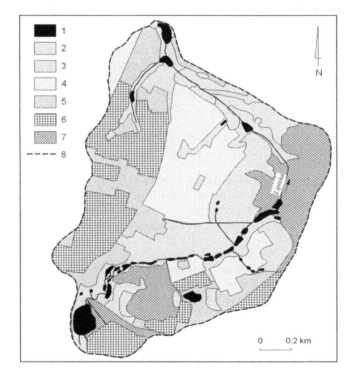

Figure 2.   Land use of the catchment of the Lipinka River. Key: 1 – anthropogenic water reservoir and wetlands, 2 – meadow, 3 – forest, 4 – arable land, 5 – gardens, 6 – build-up area, 7 – waste dump, 8 – water shed of the catchment area.

18°53'43.41"E. The surface area of the catchment is 2.94 km² while the length of the Lipinka River is 3.6 km. The land use pattern in the catchment area is shown in Fig. 2. At the turn of the 19th and 20th c., there were 26 reservoirs on the Lipinka River. Most of them were used for fish farming. Currently, most of these reservoirs are disused. A lack of maintenance caused the reservoirs to

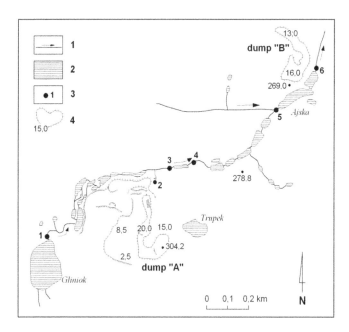

Figure 3. Location of the sampling points. Key: 1 – surface streams, 2 – anthropogenic water reservoirs and wetlands, 3 – sampling point numbers, 4 – dump borderline and a dump height.

get silted up and overgrown with water plants. The old reservoirs turned into wetlands. None of these wetlands is a purposefully designed object. The transformation into wetlands resulted from the natural spontaneous succession.

In the catchment area being discussed, there are two industrial landfills that affect the waters of the Lipinka River (Fig. 3). Landfill A contains slag from the metallurgical processing of zinc-lead ores and the waste rock from coal mines (sandstones, siltstones, shales). These two types of materials are partially mixed. This landfill has been reclaimed. The slopes of the spoil heap were covered with a layer of soil and seeded with grass. The mixing of the slag with the waste rocks created good conditions for the infiltration of rainwater into the dump. Therefore, the leachate flows out at the base of the heap (measurement site No 2) (Fig. 3). The yield of the leachate is from 0.02 to 0.1 L/s. Spoil heap B contains only slag from the metallurgical processing of zinc-lead ores. The low permeability of this type of waste reduces water infiltration. There was no discharge of the leachate from the landfill. Landfill B is directly adjacent to the shoreline of the Ajska reservoir (Fig. 3). This reservoir, despite a high level of contamination of the sediment by heavy metals (Table 1), is used for fish farming and the fish that are caught, mainly carp (*Cyprinus carpio*), are consumed by humans.

3 RESEARCH METHODS

Hydrographic mapping to assess the changes in the water conditions in the catchment area of the Lipinka River was conducted in accordance with the guidelines given by Gutry-Korycka & Werner-Więckowska (1996). Six points of water sampling for physico-chemical analyses were located on the Lipinka River (Fig. 3). Measurement site No 1 (the outflow from the Gliniok reservoir) was the control object. It is located outside the influence of the landfill. Measurement site No 2 was the leachate from the industrial waste landfill. The remaining measurement points (No's 3–6) were located on the inflow and outflow from the water reservoirs (anthropogenic wetlands). Sixteen water samples were collected at the individual measurement points. Sampling was performed at

Table 1. Concentration of heavy metals in sediments [mg/kg dry weight].

Points of sampling	Zn	Pb	Cd	Cu
1	18,700	850	93	99
2	30,500	1130	280	144
3	30,350	917	186	122
4	42,800	705	510	100
5	11,700	895	69	113
6	28,600	1690	45	79

monthly intervals (January – December 2011 years – 12 samples) and four samples in March, July, October and December 2012.

Measurements of pH, temperature and electrolytic conductivity were made in the field using a multiparameter probe EDS 6600 produced by YSI. The probe was calibrated using standard solutions before each testing. Water samples for chemical analyses were collected in polyethylene bottles. The water samples were transported at the temperature of +4°C for the laboratory tests. The samples were filtered on the 0.45 μm filter (Millipore) and acidified before the analyses.

Marking of the selected metals (Zn, Pb, Cd, Cu) was performed using an atomic absorption spectrometer of the Solar M6 type with a graphite tube with flameless atomisation. Additionally, at measurement sites No 3 and 4, the flow rate (Q) of the Lipinka River was measured. These measurements were conducted using a RBC flow valve by Eijkelkamp. The measurements were performed in the characteristic times of the year – in the spring (March), summer (July), autumn (October) and winter (December) in 2011 and 2012 (eight measurements).

The measurements of the flow rate permitted the calculation of the ion runoff according to the following formula 1:

$$A_s = T \cdot Q \tag{1}$$

where: A_s – ion outflow [mg/s]; T – ion mass [mg/L]; Q – flow rate [L/s].

To test the significance of differences the non-parametric equivalent of the one-way analysis of variance of Kruskal-Wallis was applied, while to test for multiple comparisons – the Conover test was used. All the data were presented using box-and-whiskers plots. When compared to the significance of the median differences of the variables, such as the physical or chemical parameters of the water from different objects, these differences were indicated by appropriate small letters (a, b, c) placed at the top of the figure. Different letters indicate that the values differ significantly at $p < 0.05$.

4 RESEARCH RESULTS

The effect of the leachate on the physical properties of the Lipinka River's water is best reflected by changes in the electrolytic conductivity. The average value of the electrolytic conductivity of the water of the control object is 714 μS/cm (Fig. 4). Below the inflow of the leachate, the electrolytic conductivity value increases to 5062 μS/cm (Fig. 4). The largest values of the electrolytic conductivity are characteristic of the leachate itself – 7644 μS/cm.

In addition to the high electrolytic conductivity, the leachate also contains high concentrations of heavy metals. This is best demonstrated in the case of zinc. In the control object, the average concentration of zinc is 0.29 mg/L (Fig. 5), while its average concentration in the leachate is 8.1 mg/L and the maximum concentration reached a value of 12.1 mg/L. The inflow of the leachate into the Lipinka River contaminates the water with this metal (Fig. 5). The recorded zinc concentration

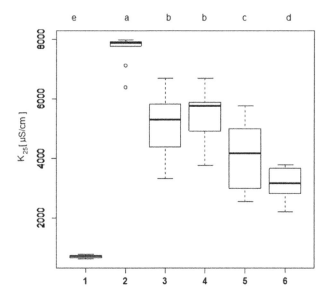

Figure 4.   Electric conductivity (n = 16).

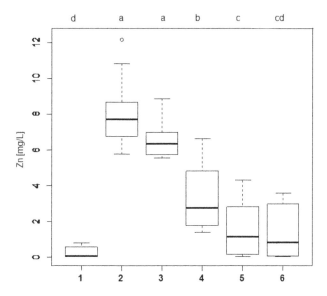

Figure 5.   Zinc concentrations (n = 16).

is many times greater than that found in the waters of the uncontaminated rivers. According to Moore (1984) zinc in the waters of uncontaminated rivers is in the range of 0.005–0.015 mg/L. The observed concentration of zinc should, therefore, be considered to be very high. A higher level of contamination (43.1 mg/L) was found by Pasternak (1974) in river water below the discharge of a zinc smelter. The flow of the water through a reservoir covered with willow moss (*Fontinalis antipyretica*) significantly affects the decrease in the concentration of zinc in the river water. Statistically significant differences in the concentrations of zinc were recorded between the water flowing into the reservoir (measurement site No 3) and the outflow (measurement site No 4) (Fig. 5). In addition to the decrease in the concentration, the reservoir also significantly reduces the load of

Table 2. Average (n = 2) removal efficiency [%] of the zinc load in the reservoir with willow moss.

Season	Sample point	Q [dm³/s]	Zn [mg/L]	A$_s$ [mg/s]	Removal efficiency [%]
spring	3	2.8	3.91	1.09	
	4	2.8	1.71	0.47	57
summer	3	1.7	8.87	1.5	
	4	1.6	1.85	0.29	81
autumn	3	2.2	5.7	1.2	
	4	2.2	1.3	0.28	77
winter	3	1.5	3.9	0.58	
	4	1.6	1.7	0.27	53

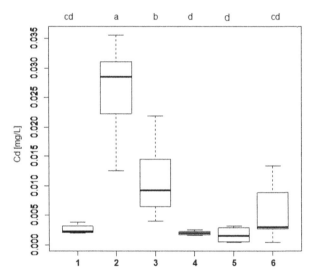

Figure 6. Cadmium concentrations (n = 16).

zinc (Table 2). The reservoir holds about 80% of the zinc load in spring and summer. A lower efficiency of 54% was found in winter.

No effect of landfill B on the increase in the concentration of zinc in the waters of the Lipinka River was observed (Fig. 5). This may be linked to the lack of the outflow of the leachate from this landfill. The only outflow of the water from the landfill is surface runoff that occurs during rainfall.

Zinc ores often contain relatively high concentrations of cadmium – approximately 5% – and therefore, the production of zinc may contaminate the aquatic environment with this metal (Alloway & Ayres 1999). This is also the case of the catchment that was studied. The leachate is characterised by very high levels of cadmium – 0.062 mg/L (Fig. 6). The inflow of the leachate into the Lipinka River causes significant contamination of its waters with this metal (Fig. 6). The degree of contamination can be observed by comparing it to uncontaminated water. The concentration of cadmium in the waters of uncontaminated rivers is 0.00002 mg/L (Kabata-Pendias & Pendias 1999). A high level of contamination with cadmium (0.017 mg/L) was also found in the water of the Sztoła River into which the mine waters from the zinc-lead mines are discharged (Świderska-Bróż 1993).

As in the case of zinc, the flow of the water through the reservoir significantly affects the decrease in the concentration of cadmium in the river (Fig. 6). There is also a very large reduction in the

Table 3. Average (n = 2) removal efficiency [%] of the cadmium load in the reservoir with willow moss.

Season	Sample point	Q [dm³/s]	Cd [mg/L]	A_s [mg/s]	Removal efficiency [%]
spring	3	2.8	0.0219	0.0613	
	4	2.8	0.0021	0.0058	90
summer	3	1.7	0.0092	0.0156	
	4	1.6	0.0019	0.003	81
autumn	3	2.2	0.0104	0.0228	
	4	2.2	0.0029	0.0063	77
winter	3	1.5	0.0074	0.0111	
	4	1.6	0.0025	0.004	64

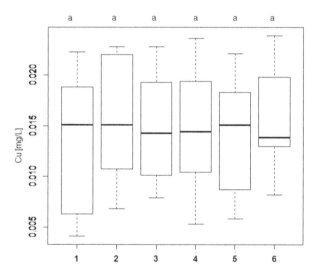

Figure 7. Copper concentrations (n = 16).

cadmium load. The greatest reduction of over 90% was found in the summer (Table 3). The high efficiency in the removal of cadmium may also be proved by a low variation in the concentration of this metal in the water flowing out of the reservoir. The value of the coefficient of variation (C_V) is 5% and is the lowest of all the measurement sites.

The average concentrations of cuprum at all of the measurement sites were similar and amounted to ≈0.015 mg/L (Fig. 7). This value does not vary from those commonly found in the waters of Polish rivers (Doijlido 1995). There were no statistically significant variations in the concentration of cuprum between the measurement sites (Fig. 7). Additionally, the concentration variation was comparable at all of the measurement points and was ≈5%. Based on these results, it can be concluded that the main source of cuprum is atmospheric pollution. The catchment of the Lipinka River is located in an area with one of the highest levels of air pollution with this metal in Poland (Chławiczka 2008).

Similar to the case of copper, there were no statistically significant variations in the concentration of lead between the measurement sites (Fig. 8). The average concentration of this metal in the water of the Lipinka River was ≈0.015 mg/L and was comparable to that which is commonly found in rivers with a moderate degree of contamination [19]. It was, however, higher than the concentration of this metal in the waters of the uncontaminated rivers, which is ≈0.002 mg/L (Brugmann 1981).

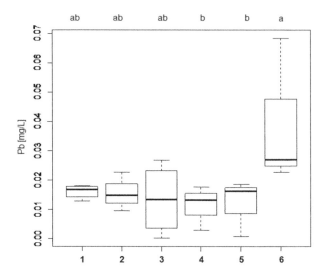

Figure 8.   Lead concentrations (n = 16).

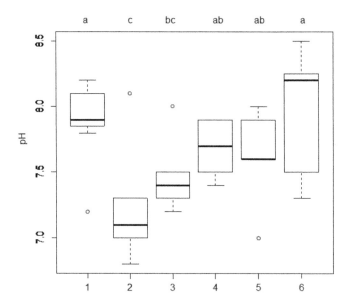

Figure 9.   Water reaction [pH] (n = 16).

A concentration higher than 1.0 mg/L was found in the water of the Biała Przemsza and Sztoła Rivers into which the mine waters from the zinc-lead mines are discharged (Świderska-Bróż 1993).

The average concentration of lead in the leachate was 0.015 mg/L and was lower than in the water in the control object (0.016 mg/L). The relatively low concentration of lead in the leachate is due to the fact that its compounds are poorly soluble at a pH close to neutral. Such pH values are characteristic of the leachate (Fig. 8). The highest average concentration of lead – 0.039 mg/L – was found at measurement site No 5 (Fig. 8). There is also a large variation in the concentration at that site (C_v – 64%). The high variations of the concentration levels were also found at measurement site No 3 (C_v – 89%). This may indicate that a migration of lead from the landfills takes place during rainfall or snowmelt. Contact of the acidic rainwater or snowmelt with the waste promotes the migration of lead. This contact is especially easy in the case of the landfill B because it is devoid of vegetation.

5 DISCUSSION

The results of the research indicate that the leachate from landfill A is an important source of water contamination. The leachate shows high electrolytic conductivity values. Such high electrolytic conductivity in this type of landfill leachate was also reported by other authors (Molenda 2006, Chmura & Molenda 2012). In the case of the leachate that was tested, the high value of the electrolytic conductivity is caused by the presence of sulphate ions (SO_4^{2-} – 3500 mg/L), sodium (Na^+ – 650 mg/L) and chloride (Cl^- – 630 mg/L) (Jankowski et al. 2006, Molenda 2006).

This confirms the research of other authors that industrial landfills are a permanent source of hydrosphere contamination (Twardowska 1981, Szczepańska & Twardowska 1999, Szczepańska & Twardowska 2004, Stefaniak & Twardowska 2006, Molenda 2006, Molenda & Chmura 2007).

The leachate of the landfill that was studied has a high concentration of heavy metals, mainly zinc and cadmium. This is associated with the significant amounts of these elements in the waste, which is confirmed by the mineralogical studies of this type of waste carried out by (Cabała 2005, Cabała et al. 2007, Cabała 2009, Jonczy 2006).

It was observed that the flow of the water through the reservoir covered with willow moss leads to a significant reduction in the load of zinc and cadmium that are transported by the river. The role of wetlands in water purification from heavy metals is known (Odurn et al. 2000, Leady & Gottgens 2001, Wojciechowska & Swarna 2006, Muthukrishnan & Swarna 2006). A similar efficiency in the removal of zinc like in the test object was also found in a natural swamp that is used for the treatment of mine water from zinc-lead ores in southern Poland (Wójcik & Wójcik 1991, Wójcik & Leszczyński 1993). A partial removal of zinc was also found in a natural swamp in Westralian Sands in Australia (Dunbabin & Bowmer 1992).

The high efficiency of the zinc and cadmium removal may be associated with the presence of willow moss in the reservoir. It should be noted that willow moss has been widely recognised as a bioindicator species of water contamination (Vazquez et al. 2004, Diaz et al. 2012). This plant has the capacity to accumulate large amounts of trace elements, including heavy metals, in a short period of time (a few days or weeks), while the period of their release is much longer (several months) (Cenci 2000). Willow moss may produce very large and dense clumps of shoots from 50 to 90 cm long, which results in a high biomass. This, in turn, affects the efficiency of capturing heavy metals from the water (Siebert et al. 1996). This moss species, due to its very thick leaves, also stops the transported slurry. In this study, the leaves of the moss specimens collected were completely covered with mineral deposits. The studies indicate that cadmium is strongly adsorbed by the suspension (Filgueiras et al. 2004, van Loon & Duffy 2007). At a pH of more than 7, almost all of this metal's ions are subject to sorption. Such a reaction was found in the water of the reservoir (Fig. 9). The cadmium absorbed by the suspension is stopped in the reservoir. In this case, willow moss has a mechanical function that consists of filtering the water from the slurry. The removal of cadmium by the sedimentation of suspensions may also be indicated by that metal's concentrations in the sediments. The highest concentration of cadmium (510 mg/kg) was found in the sediments of the reservoir with willow moss. Moreover, the concentration of zinc was the highest and amounted to 42,800 mg/kg (Table 1). A similar level of sediment contamination with heavy metals was found by Molenda (2001) in the vicinity of other metallurgical waste landfills. Given the activity of willow moss throughout the year, it has an advantage over other plants that are commonly used in hydrophytic wastewater treatment plants that are active only during the growing season, such as the common reed *Phragmites australis* (Rzętala et al. 2011). However, in the winter season removing heavy metals from the water was less efficient. Seasonal fluctuations are characteristic of this type of ecosystem (Shomar et al. 2005).

Most likely, the removal of heavy metals could be higher if the retention time of the water in the reservoir was longer. Currently, it is about two days. Under technical test conditions, the retention time of the leachate is much longer and amounts to from five to ten days (Obarska-Pempowiak 2002, Wojciechowska & Waara 2011).

6 CONCLUSIONS

The studies allow the following conclusions to be drawn:

- the leachate from landfill A is characterised by a high electrolytic conductivity and a high concentration of zinc and cadmium ions,
- the leachate inflow into the Lipinka River causes an increase in electrolytic conductivity and the contamination of the water with zinc and cadmium,
- the reservoir that is overgrown with willow moss shows a high efficiency in decreasing the concentrations and reducing the loads of zinc and cadmium,
- willow moss can be used in the hydrophytic treatment of industrial wastewater containing high concentrations of zinc and cadmium.

REFERENCES

Alloway, B.J. Ayres, D.C. 1999. Chemiczne podstawy zanieczyszczania środowiska. *Wydawnictwo Naukowe PWN* (pp. 423). Warszawa.

Brugmann, L. 1981. Heavy metals in the Baltic Sea. In The State of the Baltic, Ed. By G. Kullenberg, *Mar. Pollut. Bull*, 12(6) 214–218.

Cabała, J. 2005. Kwaśny drenaż odpadów poflotacyjnych rud Zn-Pb; zmiany składu mineralnego w strefach ryzosferowych rozwiniętych na składowiskach. *Zeszyty Naukowe Politechniki Śląskiej. Górnictwo*, 267, 63–70.

Cabała, J. 2009. Metale ciężkie w środowisku glebowym olkuskiego rejonu eksploatacji rud Zn-Pb. *Wydawnictwo Uniwersytetu Śląskiego* (pp. 141–142).

Cabała, J., Idziak, A., Kondracka M., Kleczka M. 2007. Fizykochemiczne własności odpadów występujących w obszarach historycznej przeróbki rud Zn-Pb. *Kwartalnik GIG. Wyd. Spec.*, 3, 141–152.

Cenci, R.M. 2000. The use of aquatic moss (*Fontinalis antipyretica*) as monitor of contamination in standing and running waters: limits and advantages. *J. Limnol.*, 60(1), 53–61.

Chławiczka, A. 2008. Metale ciężkie w środowisku. *Wydawnictwo Ekonomia i Środowisko* (pp. 228). Białystok.

Chmura, D., Molenda, T. 2012. Influence of thermally polluted water on the growth of helophytes in the vicinity of a colliery waste tip. *Water, Air & Soil Pollution*, 223 (9).

Diaz, S., Villares, R., Carballeira, A. 2012. Uptake Kinetics of As, Hg, Sb, and Se in the Aquatic Moss *Fontinalis antipyretica* Hedw. *Water Air And Soil Pollution*, 223 (6), 3409–3423.

Dojlido, J. 1995. Chemia wód powierzchniowych, *Wydawnictwo Ekonomia i Środowisko* (pp. 342). Białystok.

Dunbabin, J., Bowmer, K.H. 1992. Potential use of constructed wetland for treatment of industrial wastewater containing metals. *The science of the Total Environment*, 3, 151–168.

Filgueiras, A.V., Lavilla, I., Bendicho, C. 2004. Evaluation of distribution, mobility and binding behavior of heavy metals in surficial sediments of Louro River (Galicia, Spain) using chemometric analysis: a case study. *Science of the Total Environment* 330 (1–3), 115–129.

Gutry-Korycka, M., Werner-Więckowska, H. 1996. Przewodnik do hydrograficznych badań terenowych. *Państwowe Wydawnictwo Naukowe* (pp. 275) Warszawa.

Jankowski, A.T., Molenda, T., Miler, E. 2006. The role of anthropogenic wetlands in purifying leachates within industial waste deposits. *Polish Journal of Environmental Studies*, 15 (5d), 665–669.

Jonczy, I. 2006. Charakterystyka mineralogiczno-chemiczna zwałowiska odpadów produkcyjnych huty cynku i ołowiu w Rudzie Śląskiej-wirku oraz jego wpływ na środowisko. *Wydawnictwo Politechniki Śląskiej* (pp. 85) Gliwice.

Kabata-Pendias, A., Pendias, H. 1999. Biogeochemia pierwiastków śladowych. *Wydawnictwo Naukowe PWN* (pp. 398). Warszawa.

Leady, B.S., Gottgens, J.F. 2001. Mercury accumulation in sediment cores and along food chains in two regions of the Brazilian Pantanal. *Wetlands Ecology and Management*, 9, 349–361.

Molenda, T. 2001. Heavy metals in bottom deposits of anthropogenic water reservoirs in Katowice. *Limnological Review* No 1, 65–69.

Molenda, T. 2006. Charakterystyka hydrograficzno-hydrochemiczna wypływów wód odciekowych wybranych składowisk odpadów przemysłowych. *Zeszyty Naukowe Politechniki Śląskiej*, 272, 95–103.

Molenda, T., Chmura, D. 2007. Impact of industrial waste site on soils and vegetation of a river valley – case study of the colliery spoil heap "Halemba" in Katowice Upland (southern Poland). *Polish Journal of Environmental Studies*, 16(2A), 288–291.

Molenda, T., Chmura, D. 2012. Effect of industrial waste dumps on the quality of river water. *Ecological Chemistry and Engineering A.*, 19(8), 931–938.

Moor, J.A., Ramamoorthy, J. 1984. Heavy Metals in Natural Waters. *Springer-Varlag*, Berlin.

Moor, J.A., Skarda, S.M. 1992. Wetland treatment of pulp mill wastewater. *Wat. Sci. and Technol.*, 35(55), 241–247.

Muthukrishnan, S. 2006. Treatment Of Heavy Metals In Stormwater Runoff Using Wet Pond And Wetland Mesocosms, Proceedings of the Annual International Conference on Soils. Sediments, *Water and Energy*, 11 (9). http://scholarworks.umass.edu/soilsproceedings/vol11/iss1/9.

Obarska-Pempowiak, H. 2002. Oczyszczalnie hydrofitowe. *Wydawnictwo Politechniki Gdańskiej* (pp. 214). Gdańsk.

Odum, H.T., Wojcik, W., Pritchard, L.Jr., Ton, S., Delfino, J.J., Wojcik, M., Leszczyński S., Patel, J.D., Doherty, S.J., Stasik, J. 2000. Heavy Metals in the Environment Using Wetlands for Their Removal. Center for Environmental Policy and Center for Wetlands. *Environmental Engineering Sciences.* University of Florida. Gainesville, Lewis Publishers, Boca, Raton, London, New York, Washington, D.C., USA.

Pasternak, K. 1974. The influence of the pollution of a zinc plant at Miasteczko Śląskie on the content of microelements in the environment of surface waters. *Acta Univ. Wratisl.*, 1702, 85–95.

Rzętala, M.A., Rahmonov, O., Jagus, A., Rahmonov, M., Rzetala, M., Machowski, R. (2011). Occurrence of Chemical Elements in Common Reeds (*Phragmites Australis*) as Indicator of Environmental Conditions. Research Journal Of Chemistry And Environment, 15(2), 610–616.

Shomar, B.H., Muller, G., Yahya A. 2005. Geochemical Characterization of Soil and Water From a Wastewater Treatment Plant in Gaza. *Soil & Sediment Contamination*, 14, 309–327. DOI: 10.1080/15320380590954042.

Siebert, A., Bruns, I., Krauss, G.-J., Miersch, J., Market, B. 1996. The use of the aquatic moss *Fontinalis antipyretica* L. ex Hedw. as a bioindicator for heavy metals. *Sci. Total Environ*, 177, 137–144.

Stefaniak, S., Twardowska, I. 2006. Przemiany chemiczne w odpadach górniczych na przykładzie zwałowiska w Czerwionce-Leszczynach. *Górnictwo i Geologia*, 1, 77–87.

Szczepańska, J., Twardowska, I. 1999. Distribution and environment al impast of coal- mining wastes In Upper Silesia, *Poland. Environ. Geol.*, 38(3), 249–258.

Szczepańska, J., Twardowska, I. 2004. Mining waste. In Solid Waste: Assessment, Monitoring and Remediation, *Elsevier* (pp. 319–386). Amsterdam.

Świderska-Bróż, M. 1993. Mikrozanieczyszczenia w środowisku wodnym, *Politechnika Wrocławska* (pp. 144). Wrocław.

Twardowska, I. 1981. Mechanizm i dynamika ługowania odpadów karbońskich na zwałowiskach. Polska Akademia Nauk. Instytut Podstaw Inżynierii Środowiska. Zakład Narodowy im. Ossolińskich. Wydawnictwo PAN. Prace i studia, 25 van Loon, G.W., Duffy, S.J., (2007). *Chemia środowiskowa. Wydawnictwo Naukowe PWN* (pp. 614). Warszawa.

Vazquez, M.D., Wappelhorst, O., Markert, B. 2004. Determination of 28 elements in aquatic moss *Fontinalis antipyretica* Hedw. and water from the upper reaches of the River Nysa (Cz, D), by ICP-MS, ICP-OES and AAS. *Water Air And Soil Pollution*, 152(1–4), 153–172.

Wojciechowska, E., Waara, S. 2011. Distribution and removal efficiency of heavy metals in two constructed wetlands treating landfill leachate, *Water Sci. Technol.* 64, 1597–1606.

Wójcik, W., Leszczyński, S. 1993. Economical aspects of mine wastewater treatment on wetland, in: L. Dzienis (Ed.), *Problemy gospodarki wodno-ściekowej w regionach rolniczo- przemysłowych* (pp. 58–70). Białystok.

Wójcik, W., Wójcik, M. 1991. Heavy metals interactions with the Biala River wetland Part I and Part II. *The University of Florida Centre for Wetlands*, (pp. 112) (part I), (pp. 126) (part II).

Progress in Environmental Engineering – Tomaszek & Koszelnik (eds)
© 2015 Taylor & Francis Group, London, ISBN: 978-1-138-02799-2

The possibilities of limitation and elimination of activated sludge bulking

M. Kida, A. Masłoń, J.A. Tomaszek & P. Koszelnik
Department of Chemistry and Environmental Engineering, Faculty of Civil and
Environmental Engineering, Rzeszów University of Technology, Rzeszów, Poland

ABSTRACT: This paper focuses on presenting causes of activated sludge bulking, as well as methods that can prevent or eliminate this disadvantageous phenomenon. In some cases, technological and operational treatments are sufficient, however, frequently the usage of appropriate chemicals, which have a negative impact on the biological activity of the activated sludge, is necessary. For this reason, alternatives, such as usage or addition of bio-specimens or airborne substances, is more frequently being sought.

1 INTRODUCTION

Most commonly used method for biological wastewater treatment is the activated sludge method. This process is modelled on a naturally occurring phenomenon in surface waters: self-purification. The basics of it involve mineralization of organic pollutants by microorganisms of activated sludge. The prerequisite for ensuring the effectiveness of the process, in addition to acquisition of the dissolved and colloidal organic matter through biocenosis of activated sludge in a form of a flocculent suspension, is mostly its separation from the treated sewage in the sedimentation stage of the process. The intensity of the treatment can be enhanced by providing the best possible contact of sewage substance with the microorganism's biomass, but even the activated sludge that removes nutrients efficiently, that is difficult in separation from its supernatant liquid, during the process of sedimentation, is inconvenient from the point of view of active sludge technology in wastewater treatment (Drzewicki 2005, Miksch & Sikora 2010, Dymaczewski 2011, Sołtysik et al. 2011). Bad sedimentation of sludge can be caused by a number of different effects such as dispersion increase, foaming, non-filamentous bulking, filamentous bulking and exuding sludge. These phenomena appear with varying intensity and are dependent on the type of technology used in wastewater treatment and its parameters, as well as the composition of the raw sewage material and the seasons of the year with their associated weather conditions (Traczewska 1997, Drzewicki 2007).

Issues related to activated sludge bulking are the subject of numerous research and developments. This research includes both, the composition of activated sludge, and technological process conditions related to the formation of this phenomenon, as well as the ways to prevent and limit sludge bulking (Traczewska 1997, Eikelboom & Grovenstein 1998, Gerardi 2002, Jenkins et al. 2004, Kocwa-Haluch & Woźniakiewicz 2011, Lemmer 2000, Masłoń et al. 2012). This paper is a review of methods used in the context of limiting and preventing bulking of activated sludge in the process of wastewater treatment.

2 REASONS FOR BULKING SLUDGE FORMATION

Bulking sludge is defined as active sludge with insufficient sedimentation properties and compact abilities, caused by the excessive development of specific filamentous organisms

Table 1. The physiological differences between floc-forming bacteria and filamentous bacteria (Traczewska 1997).

Characteristics	Floc-forming bacteria	Flossing bacteria
The ability to the maximum binding of the substrate	high	low
The ability of the maximum specific growth	high	low
Endogenous ability to dying	high	low
Lowering the specific ability to grow in the presence of low concentrations of substrate	sensitive	moderate
Resistance to starvation	low	high
Loss of the ability to specifically increase at low dissolved oxygen content	sensitive	moderate
The ability to sorption of organic compounds by their excessive access	high	low
The ability to use nitrate as an electron acceptor	influence	lack
The ability to bind phosphorus	influence	lack

(Eikelboom & Grovenstein 1998, Lemmer 2000, Gerardi 2002). Bulking of activated sludge is a common problem occurring in biological wastewater treatment plants around the world. Most common causes for this problem to occur include the excessive growth of certain strains of filamentous bacteria. Filamentous bacteria are the major part of biocenosis of activated sludge. They are usually microorganisms with rod or spherical shapes that multiply by division, followed by their separation. They form a structure to which the floc-forming bacteria adhere. Their excessive growth leads to sedimentary disruption o sludge, foam development and sludge flotation. The domination of filamentous bacteria over floc-forming bacteria results primarily in their distinctive physiology (Table 1).

The phenomenon of sludge bulking is also caused by an excessive production of exopolysaccharides by zooglean bacteria. Consequently, a decrease in floc content and their ability to sediment occurs. This type of bulking, called the non-filamentous (zooglean) bulking, is, however, very rare and can be corrected by chlorination (Traczewska 1997, Wójcik-Szwedzińska 2003, Nielsen et al. 2009, Fiałkowska et al. 2010, Sołtysik et al. 2011). Among the approximately 30 species of filamentous bacteria identified in activated sludge, for more than 90% of the sedimentary problems, only 10 types of filamentous microorganisms are responsible. (Table 2) (Lemmer 2000, Walczak et al. 2005, Bazeli 2005, 2008, 2009, Kocwa-Haluch & Woźniakiewicz 2011). There are basically three groups of filamentous bacteria: (1) sulfur bacteria – types: *Thiothrix and Beggiatoa* and 0914, 021N type, (2) Gram-negative bacteria – *Sphaerotilus* sp, Type 1701 and 1863, *Haliscomenobacter hydrossis*, (3) Gram-positive bacteria – *Actinomycet*es such as *Microthrix parvicella*, type 1851 and *Nostocoida limicola* (Lemmer 2000). The type and contribution of most frequently occurring filamentous bacteria in bulking sludge results in the type of sewage (Table 3), as well as environmental and climate conditions (Lemmer 2000, Krhutková et al. 2002, Horan et al. 2004, Kocwa-Haluch & Woźniakiewicz 2011). In Poland, the most predominant strains causing operational problems of activated sludge are *Microthrix parvicella, Type 021N* and *Actinomycetes* (Bazeli 2005).

The development of an excessive amount of specific groups of filamentous bacteria might be an indicator of the specific conditions occurring in bioreactors. Therefore, the reason for the massive development of these organisms may include low system load, low concentration of nutrients, low concentration of dissolved oxygen, high BOD5 sludge load variation, increased presence of toxic compounds as well as presence of reduced sulphur forms and fatty acids in the inflowing sewage and pH < 6.5 in the chamber. Sewage with a high proportion of carbohydrates is prone to the ability of bulking. In relation to this, the absence or presence of specified strains of microorganisms provides information about the operation of the facility (Table 4) (Traczewska 1997, Wójcik-Szwedzińska 2003, Drzewicki 2007, Dymaczewski 2011).

The phenomenon of sludge bulking is observed periodically with varying intensity, over the year (Krhutková et al. 2002). In many municipal wastewater treatment plants, bulking sludge occurs on

Table 2. Main filamentous microorganisms present in the sludge swollen (Bazeli 2005).

Type	Share (%)
Type 021N	23,3
Microthrix parvicella	15,2
Type 0041	14,6
Spaerotilus natans	9,0
Nocardia sp.	7,3
Haliscomenbacter hydrosis	4,8
Nostocoida limicola	4,2
Type 1701	3,4
Type 0961	2,8
Type 0803	2,5

Table 3. Effect of sewage origin in the dominance of filamentous bacteria (Bazeli 2008).

Sewage source	The dominant filamentous bacteria
municipal sewage	*Microthrix parvicella, Actinomycetes, Type 0041*
meat industry	*Type 021N, Thiothrix sp., Type 0041, Nostocoida limicola*
distillery	*Type 021N, Typ 0041, Nostocoida limicola*
fruit and vegetable industry	*Type 021N, Spaerotilus natans, Type 0041, Thiothrix sp, Beggiatoa*
brewery	*Spaerotilus natans, Type 021N, Typ 1701*
dairy	*Type 0092, Type 021N, Haliscomenobacter hydrossis, Nostocoida sp.*
pulp and paper industry	*Type 0041, Type 021N, Actinomycetes, Thiothrix sp.,*

Table 4. The presence of filamentous organisms, depending on the environmental conditions in the reactor (González 2008, Kamińska 2008).

Cause	Filamentous organism
Low concentrations of oxygen	*Spaerotilus natans, Type 1701, Haliscomenbacter hydrosis*
Low sediment load	*Type 0041, Type 0675, Type 1851, Type 0803*
Rotten sewage supply	*Type 021N, Thiothrix I i II, Nostocoida limicola I, II, III*
Supply of fatty substances	*Nocardia sp., Microthrix parvicella, Type 1863*
The deficit of nutrients	*Type 021N, Thiothrix I i II, Nostocoida limicola III, Haliscomenbacter hydrosi, Spaerotilus natans,*

a regular basis throughout the entire year, and, additionally, process of foaming during the seasons of autumn and spring appears (Bazeli 2008). Bulking and foaming of activated sludge determines mainly incoming sewage composition as well as operating conditions of activated sludge system (Traczewska 1997, Gerardi 2002, Dymaczewski 2011, Kocwa-Haluch & Woźniakiewicz 2011). Knowledge about the incoming sludge allows to perform risk assessment of the formation of bulking sludge (Fig. 1) (Comas et al. 2008).

The basic parameter, useful in determining the condition of sludge in terms of the deterioration of the active sludge sedimentation properties and in determining the degree of its bulking, is the sludge volume index introduced by Mohlman. It is assumed that a high value of the index $> 150\,cm^3 \cdot g^{-1}$ characterizes bulking or overly dispersed sludge (Eikelboom & van Buijsen 1999, Lemmer 2000, Jenkins et al. 2004). Conversion of the sludge floc population into bulking sludge can be estimated by determining the number of filamentous microorganisms, also known as FI indicator (*Filamentous Index*) (Jenkins et al. 2004). Identification test, as well as the assessment of the development dynamics of filamentous bacteria, can be estimated basing on classical microbiological methods

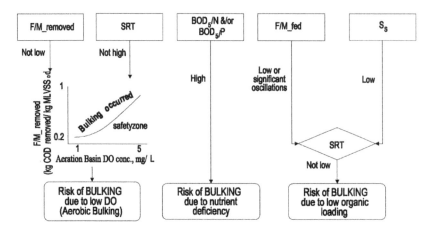

S_S – the readily degradable

Figure 1. The risk assessment of the active sludge bulking (Comas et al. 2008).

(Eikelboom & Grovenstein 1998, Gerardi 2002), molecular biology methods (Loy et al. 2002), and digital analysis of microscopic image techniques (Liwarska-Bizukojć 2005).

3 WAYS OF APPLICATION OF METHODS LIMITING AND ELIMINATING THE PROCESS OF ACTIVATED SLUDGE BULKING

Currently, operators of wastewater treatment facilities are struggling with serious problems related to the selection of appropriate courses of action in order to reduce the phenomena associated with the presence of filamentous bacteria, and therefore the choice of an appropriate strategy. Hence, the identification and determination of dominant filamentous bacteria species is the basis for an effective elimination method development for foaming and bulking of activated sludge (Lemmer 2000, Pajdak-Stós & Fialkowska 2009).

In many cases, there are processes of purification that contributes to the equilibrium, distortion, it might occur due to a low level of oxygen – required in some phases of the process for a correct denitrification or phosphorus removal procedure. Afterward, if a possibility exists, procedures reducing the maximum content of oils and fats in the wastewater should be applied before any particular reduction actions towards filamentous microorganisms will be performed. Presence of detergents should be avoided, septicity of sewage should be reduced, low-loads of organic compounds should be avoided >0.3 kg BOD $(kg/DM/day)^{-1}$, sludge age cannot be to long (<15 days) appropriate concentration levels of nutrients in wastewater, fed into the bioreactor, should be maintained (>0.2 mg P $\cdot$ dm^{-3}, >2 mg N $\cdot$ dm^{-3}) (González 2008).

Before selecting appropriate remedies for sedimentation problems related to the presence of bulking sludge, it is necessary to recognize the real causes of disruption and use microscopic examination of activated sludge, which later provides information about the physiological state of sludge flocs (Wójcik-Szwedzińska 2003).

In order to reduce the intensive development of filamentous bacteria and, consequently, prevent and neutralize sludge bulking, technology treatments should be applied first. For further development of filamentous bacteria it is essential to use other measures, such as the application of enzymatic bio-specimens or airborne substances. Despite extensive studies, universal methods preventing sludge bulking and fighting against all types of filamentous bacteria, have not yet been developed (Wójtowicz 2005, González 2008).

Table 5. Typical operational problems, their causes and methods of prevention technology (Dymaczewski 2011).

Problem	Cause	Counteraction
The presence of filamentous bacteria	low concentration of nutrients, low DO, rotten wastewater containing sulfides, filamentous bacteria present in wastewater influent, pH < 6.5 in the bioreactor	increase in intensity of aeration
high sludge index	short SRT, long SRT, supply of industrial wastewater, incorrect concentration of dissolved oxygen	modified sludge age, change in the concentration of sludge, installing baffles in the sludge chamber, reduction of longitudinal mixing of the influent wastewater and sludge at the beginning of the chamber, installation of aeration chamber before the chamber contact
A high index of sediment deposits on the surface of the particulate trap, sediment settles slowly	the presence of toxic sludge, causing the fragmentation	reducing the level of sludge recirculation
The precipitate was dispersed, presence of sludge flocs in the effluent from the settling	long SRT, low OLR, low concentration of wastewater	reduction of SRT, decrease of HRT

4 USE OF TECHNOLOGICAL AND OPERATIONAL SOLUTIONS IN THE ASPECT OF LIMITATION AND ELIMINATION OF SLUDGE BULKING

Methods based on technological treatments are included in a group of preventive methods, which main objective is to favour the growth of flocculating bacteria. It is therefore necessary to find such environmental conditions of active sludge, in which the optimal development of these bacteria would occur, due to the fact that the types of bacteria, for which the prevailing conditions are close to optimum, are dominant. This is caused by a strong competition between filamentous bacteria and floc-forming bacteria occurring in the environment of the active sludge (Pławecka 2005, Bazeli 2008).

The appearance of dark brown foam in the biological chambers shows the initial stage of an excessive multiplication of filamentous microorganisms. This conditions can be overcome by reducing the age of sludge and through increasing the oxygenation in the oxidation chambers, which will contribute to floc breaking and releasing unwanted bacteria on the surface, and if possible removing the foam from the system (Moszczyńska 2007). What appears to be an effective procedure preventing sludge bulking is the increase of the intensity of aeration, and indirectly, aeration of sewerage that supply the aeration chamber. This procedure enables to inhibit the growth of bacteria using the reduced sulphur compounds as electron donors. It is also advantageous to increase the level of external recirculation of secondary sedimentation tank. This is to ensure the reduction of sludge retention time in sedimentation tanks, which prevents the deflocculation and allows faster access to the anaerobic chamber (Bartkiewicz & Umiejewska 2004). Suitable diagnostics of operational problems arising in the wastewater treatment plant allows to conduct an optimal treatment in order to prevent the phenomenon of sludge bulking and restore the satisfactory levels of efficiency of activated sludge process (Table 5).

The application of biological selector can be included in the group of prevention methods for sludge bulking. This solution requires the use of selector or a chamber of suitable construction, in which a change in the relevant technological parameters occurs. Selectors, in which aerobic, anaerobic or anoxic conditions occur, might be used (Davoli et al. 2002, Pławecka 2005, Gray et al. 2010). The type of biological selector is conditioned by activated sludge configuration as well as the wastewater treatment plant specificity (Traczewska 1997). Application of biological selectors not always results in improved sedimentary properties of activated sludge (Gray et al. 2010). Control of dissolved oxygen concentration in biological selectors substantially determines the reduction of sludge bulking in activated sludge system (Parker et al. 2004).

Selecting the right solution depends largely on the specificity of the wastewater treatment plant. In municipal wastewater treatment plants, with industrial sewage inflow, in a number of technological treatments used to reduce sludge bulking phenomenon, elimination of the impact of industrial wastewater doses is crucial. However, in the high-performance municipal wastewater treatment plants, reducing the age of the sludge is the most efficient treatment. Activities that aim at embedding or skimming the sewage scum off are also popular. In case of industrial wastewater treatment plants, in conventional systems, an increase in the degree of aeration is planned. Using anaerobic or anoxic zones based on a metabolic selection is also preferable. This reduces the growth rate of filamentous bacteria, which, in industrial systems, are in most cases aerobic organisms. It shows that any wastewater treatment plant has to practice their own procedures to eliminate the growth rate of filamentous bacteria (Bazeli 2008, Drzewicki 2009).

5 CHEMICAL METHODS OF LIMITATION AND ELIMINATION OF BULKING SLUDGE

Controlling the population of filamentous bacteria in activated sludge can be achieved by chemical methods. The idea of chemical reagents application is based on the assumption that the filamentous microorganisms are placed on the outside of the activated sludge flocs, which makes them more sensitive to oxidizing agents in comparison to the floc-forming bacteria. Due to the physiological and morphological differences, various types of filamentous bacteria respond differently to applied chemical compounds, used to control them (Fiałkowska et al. 2008). Therefore, the choice of a compound that would limit the number of filamentous organisms should be related closely to the type of the dominant bacteria responsible for activated sludge bulking (Bazeli 2005). Improving the sedimentation properties of the activated sludge and in consequence, reducing the bulking phenomenon, may be achieved by using chemical oxidants (ozone, sodium hypochlorite, hydrogen peroxide) (Böhler M. & Siegrist H. 2004, Walczak & Cywińska 2007), aluminium salts (Roels et al. 2002, Czerwionka 2007, 2008, Drzewicki & Tomczykowska 2008, Dobiegała & Remiszewska-Skwarek 2009, Drzewicki 2009, Dominowska 2011), and iron salts Mamais et al. 2011, Skalska-Józefowicz 2007). Polyelectrolytes, which reduce the hydration level, might be used as well (Juang 2005). Using hybrid reagents, such as polyaluminum chloride, and cationic polymer (Czerwionka 2007) or a mixture of iron sulphate (III) and the anionic polymer or iron sulphate (II) modified with polymers is also possible (Moszczyńska 2007). However, the most commonly used chemical reagents, are salts of aluminium and iron, applied in the form of PIX and PAX coagulants with diverse content of Al^{3+}, Fe^{2+} and Fe^{3+}. The addition of coagulants and flocculants leads to significant improvement of the structure and growth of floc size in activated sludge (Guo et al. 2010). Chemical compounds determine the number of filamentous bacteria and macroscopic characteristics of the activated sludge in different ways (Table 6).

Knowledge of the dominant type of filamentous bacteria allows a selection of optimum chemical compound. However, usage of the same chemical reagent in the context of limiting selected filamentous organism type can provide comparable results in different configurations of the wastewater treatment plant. For example, application of polyaluminum chloride (PAX-18) in an amount of 3.0 g of Al^{3+} $(kg\,DM \cdot d)^{-1}$ in one facility, resulted in only temporary loss of foam and a sewage scum in the bioreactor (Geneja et al. 2003), while in the second one, helped to eliminate the scum completely (Dominowska 2011). In another case, a lower dose of polyaluminum chloride (PAX-16) in

Table 6. Efficacy of selected chemicals in the elimination of various filamentous bacteria (Wójtowicz 2005).

Selective agent	*M. Parvicella*	*021N*	*0041*	*Nocardia*	*Nostocoida*
PIX 113	–	+	+	?	+
PIX 200 PLUS	+/–	+	+	?	+
PAX 25	+/–	+	+	+	+/–
PAX 18	+	–	–	+	–
Fercat B (2%)	+/–	+	+	+	+
PIX + anionic polymer	+/–	+	+	+	+

+ high efficiency, +/– average efficiency, – low efficiency, ? no effect.

Table 7. Effect of chemicals to improve the sedimentation properties of activated sludge.

Type of reagent	Reagent	SVI $[cm^3 \cdot g^{-1}]$ Without reagent	With reagent	Ref.
PAX 18	$3.0 \text{ g Al}^{+3} \cdot \text{kg DM}^{-1}$	200	120	Dobiegała & Remiszewska-Skwarek 2009
PAX 16	$1.97 \text{ g Al}^{+3} \cdot \text{kg DM}^{-1}$	260	150	Geneja 2008
PAX 16	$3.0 \text{ g Al}^{+3} \cdot \text{kg DM}^{-1}$	>200	<100	Geneja 2008
FERCAT 106	$35.0–50.0 \text{ g} \cdot \text{m}^{-3}$	110	79	Moszczyńska 2007
PIX 113	$0.12 \text{ g} \cdot \text{m}^{-3}$	260–280	214	Kamińska 2008
PIX 200 PLUS	$100 \text{ g} \cdot \text{m}^{-3}$	ok. 210	<100	Wójtowicz 2005
sodium hypochlorite	$200 \text{ g} \cdot \text{m}^{-3}$	124	above 35	Walczak & Cywińska 2007
sodium hypochlorite	$100 \text{ g} \cdot \text{m}^{-3}$	124	above 45	Walczak & Cywińska 2007
H_2O_2	$200 \text{ g} \cdot \text{m}^{-3}$	124	above 40	Walczak & Cywińska 2007
H_2O_2	$100 \text{ g} \cdot \text{m}^{-3}$	124	above 45	Walczak & Cywińska 2007
Chlorine	$4–8 \text{ mg Cl}_2 \text{ g DM}^{-1} \text{ d}^{-1}$	500	<300	Ramirez et al. 2000
$FeSO_4 \cdot 7H_2O$	$30 \text{ mg Fe}^{2+} \cdot \text{dm}^{-3}$	>300	50	Agridiotis et al. 2007

an amount of 2.0–2.5 g of $Al^{3+} \cdot (\text{kg DM} \cdot \text{d})^{-1}$ ensured the elimination of foam in bioreactors and secondary settling tanks after a few days of applying this dosage (Geneja 2008).

When selecting a suitable reagent to adjust the microbial activated sludge composition, besides recording the dominant filamentous organisms occurring, the determination of microorganisms appearing in subordinated quantities is crucial. In such situation, a matter related to the selective action of a chemical reagent, used against the dominating organism, in the same time, creating possibilities for better development for subordinated organism, happens frequently. As a result of competition disappearance, the organism adopts invasive forms (Wójtowicz 2005, Walczak & Cywińska 2007). Depending on the chemical reagent applied, a different effect to improve sedimentation properties of activated sludge appear and are represented by the SVI index (Table 7).

Despite of positive effects on the reduction of filamentous bacteria and activated sludge bulking phenomenon, chemical reagents dosing, especially coagulants, can be harmful to the remaining biocenosis of the activated sludge (Lemmer 2000, Gerardi 2002). Dosage of low quantities of reactants does not impair the activity of sludge (Mamais et al. 2011), and often causes improvement in its biodiversity (Kruszelnicki 2012). Significant toxicity of chemical coagulants appear already at concentrations above $100 \text{ g} \cdot \text{m}^{-3}$ (Walczak & Cywińska 2007). The concentration of potassium permanganate and sodium hypochlorite above $175 \text{ mg} \cdot \text{dm}^{-3}$ drastically affects the microflora of activated sludge, especially organisms like ciliates and rotifers (Walczak et al. 2005). In addition to elimination of filamentous bacteria, coagulants application in membrane systems MBR (Membrane

Bioreactor) brings additional benefits. Application of chemical coagulants in such systems improves membrane functionality through the reduction of its fouling (Zhang et al. 2008, Ji et al. 2010, Huyskens et al. 2012).

Despite satisfactory effects of the coagulants used in terms of filamentous bacteria elimination, chemical treatments used are costly. The application of aluminium and iron coagulant is related to additional production of excess amount of sludge due to the precipitation of phosphates presented in wastewater. Dosing of coagulants quantity also leads to an increase in wastewater salinity and changes in pH of activated sludge. For example, addition of $100\,\text{mg} \cdot \text{dm}^{-3}$ of PIX 111 coagulant causes an increase in chlorides levels in wastewater in the amount of $26.2\,\text{mg Cl} \cdot \text{dm}^{-3}$, and PIX 113 the increase of sulphates levels in the amount of $28.8\,\text{mg SO}_4^{-2} \cdot \text{dm}^{-3}$ (Kucharski et al. 2000). Also, from industrial coagulants, harmful substances like heavy metals may be presented in the sludge, as well as in sewage due to fact that in the production of PIX and PAX waste materials (digested acids, waste metal ores and scrap) are used.

Recent professional literature reports also indicate the use of biocides to eliminate filamentous bacteria. For example, dosage of cetyltrimethylammonium bromide (CTAB) in an amount of $30\,\text{mg}$ of CTAB g DM^{-1} made a significant reduction of the occurrence of 021N filamentous bacteria, resistant to chlorination, possible (Guo et al. 2012). Recirculated sludge ozonation is another innovative solution (Lyko et al. 2012). A measurable effect of ozonation is reducing the growth of activated sludge. Dose of ozone, in the amount of $0.03–0.05\,\text{mg O}_3 \cdot \text{g DM}^{-1}$, can eliminate bulking of activated sludge as well as its growth (Chu et al. 2009).

The use of chemical substances is a part of complementary elements of the technological and operational procedures, used to reduce bulking of activated sludge. Unfortunately, chemical substances do not remove causes of sludge bulking, and the effect which can be obtained after their application is not permanent. The need for a separate tank and dosing installation is another drawback, that increases operating costs of wastewater treatment plants. Therefore, alternative solutions to replace chemicals substances are being investigated (Pławecka 2005).

6 ALTERNATIVE METHODS OF SLUDGE BULKING ELIMINATION

6.1 *Enzymatic bio-specimens in sludge bulking elimination*

One of the alternative ways of limiting and eliminating bulking sludge is the use of enzymatic bio-specimens that effect the efficiency increase of contaminant removal from sewage in activated sludge systems. The main components of bio-specimens are bacteria strains, enzymes or microorganisms, that affect the enhancement or acceleration of decomposition processes of organic compounds in wastewater, as well as nutrients and carrier media (Krajewski 2006, Matuszewska 2011).

All micro-organisms in the group of bio-specimens must belong to the first category of hazard, which include those that do not constitute a threat to the environment. They may not have a negative effect on the environment or disrupt the biological equilibrium especially in surface and ground waters and in soil (Matuszewska 2011). They are safe to use, due to the fact that they encourage natural processes for which they are genetically adapted and do not contain genetically modified micro-organisms (Grabas et al. 2010, Mendrycka & Stawarz 2012). Although bio-specimens are used in wastewater treatment technology mainly to decompose fats, remove organic pollutants (Krajewski 2006) and dispose sewage sludge (Grabas et al. 2009), scientific literature reports only a minor research studies of bio-specimens on activated sludge. Dosage of bio-specimens has important advantages: not only does it improve the efficiency of purification of wastewater, but also causes irreversible breakdown of the chemical structure of organic compounds present in the wastewater, decomposition to smaller particles with simpler structure, and then, further, the breakdown of structures by selected bacteria to H_2O and CO_2 (Krajewski 2006, Matuszewska 2011). Therefore, the dosage of bio-specimens may significantly affect the elimination of conditions favouring the development of filamentous bacteria. One of the few studies have shown that dosage of AQUATOP® BA bio-specimen leads to the formation of flocs with very good sedimentation conditions and reduces the growth of filamentous bacteria (Posavac et al. 2010). Grabas et al. (2010) have shown that EM- BIO® bio-specimen increases the efficiency of removing impurities from wastewater from the activated sludge system while not causing its bulking, and does not have a disadvantageous effect on its sedimentation parameters. In the presence of specific enzymatic-bacterial preparations activity, their use in wastewater treatment technology makes them future-attractive as they allow the decomposition of refractive, organic pollutants, fats and oils (Krajewski 2006). Recent literature reports also indicate the possibility of applying vitamins and trace metals (Co, Zn, Cu, Mo, Se), into activated sludge, in order to improve its physiological conditions, as well as to intensify the removal of pollutants from the sewage (Barnett et al. 2012).

Use of any innovative enzymatic bio-specimens in the activated sludge technology, in terms of limitation or elimination of sludge bulking phenomenon, requires an extensive knowledge of the detailed composition and type of microorganisms found in commercial products and further research.

6.2 *Natural methods of filamentous bacteria elimination from the trophic level*

Another unconventional solution in filamentous bacteria elimination from activated sludge is their natural elimination on the trophic level (Fiałkowska & Pajdak-Stós 2008, Kocerba-Soroka et al. 2013). The presence of selected strains of rotifers in activated sludge can limit the growth of filamentous bacteria effectively. Research of Fiałkowska and Pajdak-Stós (2008) have shown that rotifers of *Lecane* type feed on *Microthrix parvicella bacteria,* dramatically reducing their amount in the sludge biocenosis. Rotifers of *Lecane* type reproduce parthenogenetically, reaching a high concentration quickly, if only given the right conditions, so that they can increase in number (Pajdak-Stós & Fiałkowska 2009). In addition, Rotifers that reduce *Microthrix parvicella* bacteria, limiting the production of sludge at the same time, which has been considered a positive side effect of activated sludge bulking elimination method.

6.3 *Elimination and reduction of sludge bulking by use of mineral powdered substances*

Mineral airborne materials with a particle size $< 300\,\mu m$, depending on their physico-chemical characteristics, when dosed into sludge, may fulfil different but complementary roles of micro-carriers of biological membrane, sorbent of chemical substances and flock weight-sinkers (Masłoń & Tomaszek 2012).

The presence of powdered substances affects the morphological characteristics of activated sludge, particularly the shape and appearance, structure and consistency and size of floc, since the powdered particles are integrated with activated sludge. Probably there is an interaction between the structures of powdered particles with flocs of activated sludge so that the mineral particles are very firmly incorporated into the floc and they do not degrade in the aeration chamber or in a secondary settling tank. The incorporation of particles of powdered substances into flocs of activated sludge increases its relative density (Chudoba & Pannier 1994, Cantet et al. 1996, Lemmer 2000, Masłoń & Tomaszek 2012) Dosing the activated sludge with a form of powdered mineral substances leads to an increase of its density, resulting in an increase of sedimentation velocity and decrease of its volume index (Masłoń & Tomaszek 2012). Mineral powdered substances play a role of a weight, a sinker of activated sludge (Lemmer 2000). Issues related to activated sludge technology supported by powdered mineral substances have been widely presented in work of Masłoń and Tomaszek (2012).

As preventive actions for foaming and bulking, one might use powdered mineral substances, which, unlike chemical compounds, have no negative effect on the biological activity of activated sludge (Lemmer 2000, Masłoń & Tomaszek 2012). The quantity and type of powdered materials determine the improvement of sedimentation properties of activated sludge (Table 8).

Their tendency to incorporate in the structure of activated sludge flocs determines their usability in elimination of sludge bulking by increasing its specific gravity (Cantet et al. 1996, Eikelboom & Grovenstein 1998, Clauss et al. 1999). In authors research it is shown, that insignificant amounts of airborne substances ($<0.5\,g \cdot dm^{-3}$), in small means, have influence on sedimentation properties

Table 8. Effect of mineral powdery substance on the index of activated sludge.

| | | SVI [cm$^3 \cdot$ g^{-1}] | | |
Powdery substances	The weight amount of activated sludge in the chamber	without the addition of the weight	after the addition of the weight	Ref.
bentonite <100 μm	1.0 g $\cdot$ dm^{-3}	100	26–55	Wiszniowski et al. 2007
talc <300 μm	3.0–4.0 g $\cdot$ dm^{-3}	91–152	44–109	Cantet et al. 1996
talc	1.57–1.89 g $\cdot$ dm^{-3}	140	55	Bidault et al. 1997
talc *Aquatal®*	0.7 g $\cdot$ (g DM)$^{-1}$	350	160	Clauss et al. 1999
talc and chlorite – PE8418	0.6 g $\cdot$ (g DM)$^{-1}$	850.0	100–125	Eikelboom & Grovenstein 1998
talc	0.7 g $\cdot$ (g DM)$^{-1}$	>250	124	Agridiotis et al. 2006
talc and chlorite – *PE8418* <300 μm	1.2 g $\cdot$ (g DM)$^{-1}$	800	<200	Piirtola et al. 1999
Hungarian modified zeolite 100 μm	0.05 g $\cdot$ (g DM)$^{-1}$	128	83	Princz et al. 2003
	0.08 g $\cdot$ (g DM)$^{-1}$	128	56	
	0.1 g $\cdot$ (g DM)$^{-1}$	128	50	
zeolite, 50–200 μm	1.0 g $\cdot$ dm^{-3}	118–124	67–74	He et al. 2007

of the activated sludge, especially on related density, sedimentation and the sedimentation rate of activated sludge flocs. It is only when the content of airborne weight-sinker in activated sludge is big enough (0.75–1.0 g $\cdot$ dm^{-3}) when it is possible to determine the improvement of sedimentation characteristics, which result in a balanced sedimentation rate of activated sludge process. This is an important fact when considering the operation of secondary settling tanks and, above all, sequence batch reactors, where all disturbances in sedimentation process can lead to decrease of wastewater treatment efficiency (Masłoń et al. 2012, Masłoń et al. 2013).

The advantage of airborne substances, in comparison to chemical agents, in the context of bulking sludge elimination, is the lack of negative impact on biocenosis and physiological condition of the activated sludge. In addition, studies have shown, that the application of airborne materials leads to intensification of biochemical processes of impurities removal from wastewater – nitrification, denitrification and biological phosphorus removal (Piaskowski & Anielak 2004, Anielak & Piaskowski 2005, Anielak 2006, He et al. 2007). Among airborne materials, which have been tested and applied in the technology of activated sludge, such as diatomites, talc, bentonite, kaolin, chlorite, ash waste, the mostly understood and known in terms of supporting the technology are zeolites. Due to a significant impact on the operation of activated sludge, airborne mineral substances can be called supplements of activated sludge (Masłoń & Tomaszek 2012).

7 SUMMARY AND CONCLUSIONS

Excessive growth of filamentous bacteria in activated sludge causes significant operational problems in wastewater treatment plants all around the world. Depending on the climate, environmental conditions, type of sewage and technological characteristics of activated sludge system, for problems related to sludge bulking, different strains of filamentous bacteria can be responsible. Identification of filamentous microorganisms, as well as understanding the factors that determine their occurrence and monitoring of their number, is crucial in terms of developing specific strategies for elimination of filamentous bacteria and sludge bulking phenomenon.

In order to prevent and counteract against activated sludge bulking, firstly, all technological procedures have to be used – maintaining a constant sludge load and the concentration of dissolved

Table 9. Effect of some of the reduction and control of activated sludge swelling.

	chemical coagulants	chemical oxidants	elimination of filamentous bacteria in the trophic chain	biopreparations	powdered mineral substances
reducing the causes of swelling activated sludge	−	−	−	+++	++
elimination of filamentous bacteria	+++ / ++	+++ / ++	only *M. parvicella*	NDA	NDA
decrease SVI	+++	+++	NDA	NDA	+++
improvement of the structure of activated sludge	++	++	NDA	NDA	+++
impact on aquatic life remaining sludge	+++ (at high concentrations)	+++ (at high concentrations)	−	−	−
the improvement in the activated sludge	+ deterioration over time	+ deterioration over time	NDA	++	+++
improve the efficiency of contaminant removal	++ (only in the case of removal P-PO$_4$)	+	NDA	++ (primarily organic compounds)	+++
the costs of applying	+++ / ++	+++ / ++	+	+++	+
impact on the external environment	++ (salinity, pH change, heavy metals)	++	−	−	−

+++ very high, ++ high, + small,− no effect, NDA no data

oxygen in the bioreactor, controlling sludge age and its high degree of recirculation. Application of technological selectors, in which mixing of sewage and recalculated sludge occurs, is also possible. In case of further development of filamentous microorganisms and the deterioration of sedimentation properties of activated sludge the use of other means of prevention is necessary. Application of chemical reagents is strongly dependent on the predominant type of filamentous microorganism responsible for the activated sludge bulking. Chemical substances like ozone, sodium hypochlorite, hydrogen peroxide, and especially aluminium and iron salts, in a form of PAX and PIX coagulants determine the abundance of filamentous bacteria differently. Often, on a single wastewater treatment facility, defined strategy brings satisfactory results, and in another, effects of analogous treatments are negligible despite the fact that the problem is connected with the same filamentous bacteria types. Application of chemical coagulants, despite having high elimination effects on filamentous bacteria and sludge bulking, is costly and harmful to remaining biocenosis of activated sludge. Another drawback of coagulants application is increased value of wastewater salinity, changing the pH and enabling the possibility of heavy metals leak into the treated sewage or excessive sludge.

Increasingly, to prevent and counteract sludge bulking, alternative solutions, which include the use of airborne substances as well as enzymatic bio-specimens are applied. The advantage of the application of these methods, in comparison to chemical reagents, is that they have no negative effect on the biological activity of activated sludge, and additionally intensify individual biological processes like nitrification, denitrification, or biological phosphorus removal, and increase the efficiency of biological wastewater treatment. A relatively new solution to prevent sludge bulking is the elimination, of filamentous bacteria *Microthrix parvicella*, on the trophic level, by *Lecane* rotifers.

Despite extensive studies, conducted both in the laboratory and technical scale, a universal and effective method, that allows to control sludge bulking and eliminate all filamentous bacteria types has not yet been developed. The use of specific chemical or mineral substances as well as enzymatic bio-specimens leads, apart from the sludge bulking elimination, to significant changes in the biocenosis and physiological condition of the activated sludge, etc. All chemical, mineral and biological compounds interact differently on the biological wastewater treatment performance. Additionally, their application is also related to generation of operational costs (Table 9).

Literature related to this subject shows that despite the wide selection of chemical, mineral or biological compounds, preventive measures for the elimination of sludge bulking and filamentous bacteria should be focused, primarily, on technological activities, then on the application of environmentally friendly mineral materials, enzymatic bio-specimens, and lastly, chemical compounds.

REFERENCES

Agridiotis, V., Forster, FCIWEM C.F., Balavoine, C., Wolter, C., Carliell-Marquet C. 2006. An examination of the surface characteristics of activated sludge in relation to bulking during the treatment of paper mill wastewater. *Water and Environment Journal* 20: 141–149.

Agridiotis, V., Forster, C.F., Carliell-Marquet, C. 2007. Addition of Al and Fe salts during treatment of paper mill effluents to improve activated sludge settlement characteristics. *Bioresource Technology* 98: 2926–2934.

Anielak, A.M. 2006. Niekonwencjonalne metody usuwania substancji biogennych w bioreaktorach sekwencyjnych. *Gaz Woda i Technika Sanitarna* 2: 23–27.

Anielak, A.M. & Piaskowski, K. 2005. Influence of zeolites on kinetics and effectiveness of the process of sewage biological purification in Sequencing Batch Reactors. *Environmental Protection Engineering* 2: 31, s. 21–31.

Barnett, J., Richardson, D., Stack K., Lewis T. 2012. Addition of trace metals and vitamins for the optimization of a pulp and paper mill activated sludge wastewater treatment plant. *Appita Journal* 65 (3): 237–243.

Bartkiewicz, B. & Umiejewska, K. 2004. Zmiany w technologii oczyszczania ścieków i przeróbki osadów ściekowych. *Instal* 2: 42–46.

Bazeli, M. 2005. Wpływ wybranych koagulantów glinowych i żelazowych na dominujące bakterie nitkowate. *Seminarium Naukowo – Techniczne pt. „Nowości w zastosowaniu chemii na oczyszczalniach ścieków w oparciu o doświadczenia polskie, niemieckie i czeskie". Praga – Teplice. Materiały seminaryjne*: 38–48.

Bazeli, M. 2008. Kryteria selekcji bakterii nitkowatych w oczyszczalniach komunalnych i przemysłowych. *Seminarium Naukowo-Techniczne pt. „Nowe zastosowania chemii w technologii oczyszczania ścieków komunalnych i przemysłowych". Mikołajki, 10–12 września 2008 r. Materiały seminaryjne*: 12–19.

Bazeli, M. 2009. Rola analizy mikroskopowej w walce z bakteriami nitkowatymi. *Seminarium Naukowo-Techniczne pt. „Różne aspekty chemicznych procesów oczyszczania ścieków ze szczególnym uwzględnieniem ich energochłonności". Sopot, 30 września – 1 października 2009 r.*: 73–78.

Bidault, A., Clauss, F., Helaine, D., Balavoine, C. 1997. Floc agglomeration and structuration by a specific talc mineral composition. *Water Science and Technology* 36(4): 57–68.

Böhler, M. & Siegrist, H. 2004. Partial ozonation of activated sludge to reduce excess sludge, improve denitrification and control scumming and bulking. *Water Science and Technology* 49 (10): 41–49.

Cantet, J., Paul, E., Clauss, F. 1996. Upgrading performance of an activated sludge process through addition of talqueous powder. *Water Science and Technology* 34(5–6): 75–83.

Chu, L., Yan, S., Xing, X.-H., Sun, X., Jurcik, B. 2009. Progress and perspectives of sludge ozonation as a powerful pretreatment method for minimization of excess sludge production. *Water Research* 43 (7): 1811–1822.

Chudoba, P. & Pannier, M. 1994. Use of Powdered Clay to Upgrade Activated Sludge Process. *Environmental Technology* 15: 863–870.

Clauss, F., Balavoine, C., Hélaine, D., Martin, G., (1999). Controlling the settling of activated sludge in pulp and paper wastewater treatment plants. *Water Science and Technology* 40 (11–12): 223–229.

Comas, J., Rodríguez-Roda, I., Gernaey, K.V., Rosen, C., Jeppsson, U., Poch, M. 2008. Risk assessment model ling of microbiology-related solids separation problems in activated sludge systems. *Environmental Modelling & Software* 23: 1250–1261.

Czerwionka, K. 2007. Zastosowanie reagentów chemicznych do zwalczania bakterii nitkowatych *Nocardia*. *Seminarium Naukowo-Techniczne pt. „Różnorodność zastosowań chemii w oczyszczaniu ścieków, obróbce osadów i zwalczaniu odorów". Darłówek, 12–14 września 2007 r. Materiały seminaryjne*: 67–79.

Czerwionka, K. 2008. Zwalczanie bakterii nitkowatych typu *Nostocoida limicola I 0041* w oczyszczalni w Tłuczewie. *Seminarium Naukowo-Techniczne pt. „Nowe zastosowania chemii w technologii oczyszczania ścieków komunalnych i przemysłowych". Mikołajki, 10–12 września 2008 r. Materiały seminaryjne*: 20–27.

Davoli, D., Madoni, P., Guglielmi, L., Pergetti, M., Barilli, S., 2002. Testing the effect of selectors in the control of bulking and foaming in full scale activated – sludge plants. *Water, Science and Technology* 46(1–2): 495–498.

Dobiegała, E. & Remiszewska-Skwarek, A. 2009. Doświadczenia ze zwalczania bakterii nitkowatych w czasie modernizacji oczyszczalni ścieków "Dęgoborze". *Seminarium Naukowo-Techniczne pt. „Różne aspekty chemicznych procesów oczyszczania ścieków ze szczególnym uwzględnieniem ich energochłonności". Sopot, 30 września – 1 października 2009 r.:* 79–90.

Dominowska, M. 2011. Zastosowanie wodnego roztworu polichlorku glinu (PAX) do zwalczania bakterii nitkowatych na Oczyszczalni Ścieków "Pomorzany" w Szczecinie. *Seminarium Naukowo-Techniczne pt. „Chemia w uzdatnianiu wody i oczyszczaniu ścieków. Nowe zastosowania na bazie 20 lat doświadczeń", Szczecin, 4–7 października 2011 r. Materiały seminaryjne:* 135–143.

Drzewicki, A. 2005. Znaczenie morfologii kłaczka w procesie oczyszczania ścieków metodą osadu czynnego. *Gaz, Woda i Technika Sanitarna* 9: 26–27.

Drzewicki, A. 2007. Znaczenie obserwacji mikroskopowych osadu czynnego. *Gaz, Woda i Technika Sanitarna* 11: 22–23.

Drzewicki, A. 2009. Effect of application of polyaluminium chloride on reducing exploitation problems as the wastewater treatment plant in Olsztyn. *Polish Journal of Natural Sciences* 24 (3): 158–168.

Drzewicki, A. & Tomczykowska, M. 2008. Wpływ rozwiązań operacyjnych na ograniczenie problemów eksploatacyjnych w skali technicznej wywołanych nadmiernym rozwojem Microthrix Parvicella w osadzie czynnym. *Forum Eksploatatora* 1: 26–29.

Dymaczewski, Z. 2011. *Poradnik eksploatatora oczyszczalni ścieków.* Wyd. PZiTS. Poznań.

Eikelboom, D.H. & Grovenstein, J. 1998. Control of bulking in a full scale plant by addition of talc (PE 8418). *Water Science and Technology* 37 (4–5): 297–301.

Eikelboom, D.H. & van Buijsen, H.J.J. 1999. *Podręcznik mikroskopowego badania osadu czynnego.* Wyd. Seidel-Przywecki, Szczecin.

Fiałkowska, E. & Pajdak-Stós, A. 2008. Preliminary studies on the role of *Lecane rotifers* in activated sludge bulking control. *Water Research* 42 (10–11): 2483–2490.

Fiałkowska, E., Fyda, J., Pajdak-Stós, A., Wiąckowski K. 2010. *Osad czynny. Biologia i analiza mikroskopowa.* Wyd. Seidel – Przywecki.

Geneja, M., Czerwionka, K., Bronk, W. 2003. Chlorek poliglinu w likwidacji skutków rozwoju bakterii nitkowatych. *Przegląd Komunalny* 9 (144): 48–49.

Geneja, M. 2008. Zastosowanie glinu do ograniczania rozwoju bakterii nitkowatych w systemach osadu czynnego. *Przemysł Chemiczny* 87 (5): 452–455.

Gerardi, M.H. 2002. Settleability problems and loss solids in the activated sludge process. John Wiley & Sons, Inc., New Jersey.

González, J. 2008. Zastosowania chemii do wspomagania pracy biologii. *Seminarium Naukowo-Techniczne pt. „Nowe zastosowania chemii w technologii oczyszczania ścieków komunalnych i przemysłowych". 10–12 września 2008 r., Mikołajki:* 3–11.

Grabas, M., Czerwieniec, E., Tomaszek, J.A., Masłoń, A., Leszczyńska, J. 2009. Effectiveness of sludge processing with bio-preparation (EM-bio) and structural material. *Environment Protection Engineering* 35 (2): 131–139.

Grabas, M., Tomaszek, J.A., Czerwieniec, E., Zamorska, J., Kukuła, E., Masłoń, A., Łuczyszyn, J. 2010. Wstępne wyniki zastosowania biopreparatu EM do oczyszczania ścieków metodą osadu czynnego. *Konferencja Naukowo – Techniczna "Zaawansowane technologie biologicznego oczyszczania ścieków komunalnych". Zegrze, 21–22 kwietnia 2010 r. Materiały konferencyjne:* 83–95.

Gray, D.M., De Lange, V.P., Chien, M.H., Esquer, M.A., Shao, Y.J. 2010. Investigating the fundamental basis for selectors to improve activated sludge settling. *Water Environ. Res.* 82 (6): 541–555.

Guo, J., Peng, Y., Wang, Z. 2012. Control filamentous bulking caused by chlorine-resistant Type 021N bacteria through adding a biocide CTAB. *Water Research* 46 (19): 6531–6542.

Guo, W., Ngo, H.H., Vigneswaran, S., Dharmawan, F., Nguyen, T.T., Aryal, R. 2010. Effect of different flocculants on short-term performance of submerged membrane bioreactor. *Separation and Purification Technology* 70: 274–279.

He, S.-B., Xue, G., Kong, H.-N., Li, X. 2007. Improving the performance of sequencing batch reactor (SBR) by the addition of zeolite powder. *Journal of Hazardous Materials* 142 (1–2): 493–499.

Horan, N., Lavander, P., Cowley, E. 2004. Experience of activated-sludge bulking in the UK. *Water and Environment Journal* 18 (3): 177–182.

Huyskens, C., de Wever, H., Fovet, Y., Wegmann, U., Diels, L., Lenaerts S. 2012. Screening of novel MBR fouling reducers: Benchmarking with known fouling reducers and evaluation of their mechanism of action. *Separation and Purification Technology* 95: 49–57.

Jenkins, D., Richard, M.G., Daigger, G.T. 2004. *Manual on the causes and control of activated sludge bulking, foaming, and other solids separation problems*. 3rd Edition, IWA Publishing, London.

Ji, J., Qiu, J., Wai, N., Wong, F.S., Li, Y. 2010. Influence of organic and inorganic flocculants on physical-chemical properties of biomass and membrane-fouling rate. *Water Research* 44: 1627–1635.

Juang, D. 2005. Effects of synthetic polymer on the filamentous bacteria in activated sludge. *Bioresource Technology* 96: 31–40.

Kamińska, A. 2008. Doświadczenia w zwalczaniu bakterii nitkowatych na Oczyszczalni Zakładowej w Spółdzielni Mleczarskiej MLEKPOL w Grajewie. *Seminarium Naukowo-Techniczne pt. „Różne aspekty chemicznych procesów oczyszczania ścieków ze szczególnym uwzględnieniem ich energochłonności". Sopot, 30 września – 1 października 2009 r.*: 91–99.

Kocerba-Soroka, W., Fiałkowska, E., Pajdak-Stós, A., Klimek, B., Kowalska, E., Drzewicki, A., Salvadó, H., Fyda, J. 2013. The use of rotifers for li miting filamnetous bacteria Type 021N, a bacteria causing activated sludge bulking. *Water, Science and Technology* 67 (7): 1557–1563.

Kocwa-Haluch, R. & Woźniakiewicz, T. 2011. Analiza mikroskopowa osadu czynnego i jej rola w kontroli procesu technologicznego oczyszczania ścieków. *Czasopismo Techniczne. Środowisko* 6 (108): 141–162.

Krajewski, P., (2006). Biopreparaty – historia, rozwój i aplikacje. *Wodociągi – Kanalizacja* 1(23): 24–27.

Krhutková, O., Ruzicková, I., Wanner, J. 2002. Microbial evaluation of activated sludge and filamentous population at eight Czech nutrient removal activated sludge plants during year 2000. *Water, Science and Technology* 46 (1–2): 471–478.

Kruszelnicki, P. 2012. Walka z "nitką" – doświadczenia z czterech lat eksploatacji oczyszczalni ścieków w Lewinie Brzeskim. *Forum Eksploatatora* 4 (61): 37–39.

Kucharski, B., Kowalski, J., Uberna, A. 2000. Efekty stosowania glinianu sodowego NaAlO$_2$ w oczyszczaniu ścieków. *IX Ogólnopolskie Seminarium PZITS O/Kielce. Politechnika Warszawska, Cedzynia. Materiały konferencyjne:* 111–119.

Lemmer, H. 2000. *Przyczyny powstawania i zwalczania osadu spęczniałego*. Wyd. Seidel-Przywecki, Szczecin.

Liwarska-Bizukojć, E. 2005. Application of image analysis techniques in activated sludge wastewater treatment processes. *Biotechnology Letters* 27: 1427–1433.

Loy, A., Daims, H., Wagner, M. 2002. Activated Sludge: Molecular techniques for determining community composition, [w:] Bitton G., (ed.) *The Encyclopedia of Environmental Microbiology*. Wiley, New York, 26–43,

Lyko, S., Teichgräber, B., Kraft, A. 2012. Bulking control by low-dose ozonation of returned activated sludge in a full-scale wastewater treatment plant. *Water Science and Technology* 65 (9): 1654–1659.

Mamais, D., Kalaitzi, E., Andreadakis, A. 2011. Foaming control in activated sludge treatment plants by coagulants addition. *Global NEST Journal* 13 (3): 237–245.

Masłoń, A. & Tomaszek, J.A. 2012. Kierunki zastosowania mineralnych materiałów pylistych w technologii osadu czynnego – studium literatury. *Prace Naukowe Inżynieria Środowiska – Współczesne problemy inżynierii i ochrony środowiska* 59: 5–23.

Masłoń, A., Tomaszek, J.A., Opaliński, I. 2013. Badania nad poprawą właściwości sedymentacyjnych osadu czynnego przy zastosowaniu mineralnych substancji pylistych. *Gaz, Woda i Technika Sanitarna*, 12: 490–495.

Masłoń, A., Opaliński, I., Tomaszek, J.A. 2013. Wpływ wybranych mineralnych substancji pylistych na właściwości sedymentacyjne osadu czynnego, 186–195. [w:] Pikoń K., Stelmach S., (red) *Współczesne problemy ochrony środowiska*. Wyd. Archiwum Gospodarki Odpadami i Ochrony Środowiska, Gliwice.

Matuszewska, R. 2011. Biopreparaty w biologicznym oczyszczaniu ścieków. *Wodociągi – Kanalizacja* 11: 51–53.

Mendrycka, M. & Stawarz, M. 2012. Zastosowanie biopreparatu wspomagającego oczyszczanie ścieków metodą osadu czynnego. *Inżynieria Ekologiczna* 28: 43–56.

Miksch, K. & Sikora, J. 2010. *Biotechnologia ścieków*. Wydawnictwo PWN. Warszawa.

Moszczyńska, J. 2007. Zastosowanie koagulantów w różnych technologiach na przykładzie oczyszczalni ścieków w Lubinie. *Przegląd Komunalny* 5: 73–78.

Nielsen, P.H., Kragelund, C., Seviour, R.J., Nielsen, J.L. 2009. Identity and ecophysiology of filamentous bacteria in activated sludge. *FEMS Microbiology. Reviews* 33: 969–998.

Pajdak-Stós, A. & Fiałkowska, E. 2009. Alternatywne metody zwalczania bakterii nitkowatych w osadzie czynnym. *Forum Eksploatatora* 5 (44): 22–23.

Parker, D., Appleton, R., Bratby, J., Melcer, H. 2004. North American performance experience with anoxic and anaerobic selectors for activated sludge bulking control. *Water, Science and Technology* 50 (7): 221–228.

Piaskowski, K. & Anielak, A.M. 2004. Wpływ na osad czynny zeolitu naturalnego oraz modyfikowanego. *Inżynieria i Ochrona Środowiska* 7 (1): 39–53.

Piirtola, L., Hultman, B., Löwén, M. 1999. Activated sludge ballasting in pilot plant operation. *Water Research* 33 (13): 3026–3032.

Pławecka, M. 2005. Puchnięcie osadu czynnego – monitoring i kontrola procesu. *Forum Eksploatatora* 3: 15–18.

Posavac, S., Dragicević, T.L., Hren, M.Z. 2010. The improvement of dairy wastewater treatment efficiency by the addition of bioactivator. *Mljekarstvo* 60 (3): 198–206.

Princz, P., Oláh, J., Smith, S., Hatfield, K. 2003. Complex analytical procedure for the characterization of modified zeolite and for the assessment its effects on biological wastewater treatment. *XVII IMEKO World Congress Metrology in the 3rd Millenium, Croatia, Dubrovnik June 22–27, 2003. Proceedings*: 2118–2122.

Ramirez, G.W., Alonso, J.L., Villanueva, A., Guardino, R., Basiero, J.A., Bernecer, I., Morenilla, J.J. 2000. A rapid, direct method for assessing chlorine effect on filamentous bacteria in activated sludge. *Water Research* 34 (15): 3894–389.

Roels, T., Dauwe, F., van Damme, S., de Wild,e K.E., Roelandt, F. 2002. The influence of PAX-14 on activated sludge systems and in particular on Microtrix parvicella. *Water, Science and Technology* 46: 487–490.

Skalska-Józefowicz, K. 2007. Usuwanie fosforu i walka z pienieniem osadu przy udziale koagulantów w oczyszczalni ścieków Południe w Warszawie. *Seminarium pt. "Różnorodność zastosowań chemii w oczyszczaniu ścieków, obróbce osadów i zwalczaniu odorów". Darłówek, 12–14 września 2007 r. Materiały seminaryjne:* 25–37.

Sołtysik, D., Błaszczyk, D., Bednarek, I. 2011. Wynik obserwacji mikroskopowych osadu czynnego a zmiany parametrów fizyko – chemicznych oznaczanych w oczyszczalni ścieków. *Gaz, Woda i Technika Sanitarna* 1: 23–28.

Traczewska, T.M. 1997. Biotyczne i abiotyczne uwarunkowania pęcznienia osadu czynnego. *Ochrona Środowiska* 2 (65): 29–32.

Walczak, M., Donderski, W., Pawluk, H. 2005. Ograniczenie rozwoju bakterii nitkowatych w osadzie czynnym. *Wodociągi – Kanalizacja* 4 (13): 10–12.

Walczak, M. & Cywińska, A. 2007. Application of selected chemical compounds to limit the growth of filamentous bacteria in activated sludge. *Environmental Protection Engineering* 33 (2): 221–230.

Wiszniowski, J., Surmacz-Górska, J., Robert, D., Weber, J.V., 2007. The effect of landfill leachate composition on organics and nitrogen removal in an activated sludge system with bentonite additive. *Journal of Environmental Management* 85: 59–68.

Wójcik-Szwedzińska, M. 2003. Badania populacji bakterii nitkowatych w osadzie czynnym. *Inżynieria i Ochrona Środowiska* 6 (1): 17–26.

Wójtowicz, A. 2005. Selektywne oddziaływanie soli żelaza i glinu na mikroorganizmy nitkowate wywołujące efekty puchnięcia i pienienia na oczyszczalni ścieków w Słupsku. *Seminarium Naukowo – Techniczne pt. „Nowości w zastosowaniu chemii na oczyszczalniach ścieków w oparciu o doświadczenia polskie, niemieckie i czeskie". Praga – Teplice. Materiały seminaryjne:* 49–64.

Zhang, H.F., Sun, B.S., Zhao, X.H., Gao, Z.H. 2008. Effect of ferric chloride on fouling in membrane bioreactor. *Separation and Purification Technology* 63: 341–347.

Progress in Environmental Engineering – Tomaszek & Koszelnik (eds)
© *2015 Taylor & Francis Group, London, ISBN: 978-1-138-02799-2*

Lakes and reservoirs restoration – Short description of the chosen methods

L. Bartoszek & P. Koszelnik
*Department of Chemistry and Environmental Engineering, Faculty of Civil and
Environmental Engineering, Rzeszów University of Technology, Rzeszów, Poland*

ABSTRACT: Eutrophication of natural and man-made lakes due to high nutrient loading has been the paramount environmental problem for lakes worldwide for the past dozen years. The majority of lake restoration projects have been conducted to combat eutrophication. Lake restoration is a broad term used for different techniques aiming to bring a body of water back to or closer to anthropogenically undisturbed conditions. The paper contains description and analyse of mostly used techniques of lakes and reservoirs restoration. Methods of phosphorus inactivation in bottom sediment using Fe, Al and Ca reagents are presented mainly.

1 CAUSES OF THE ARTIFICIAL RESERVOIRS DECAYING

Artificial reservoirs are degraded and made shallow faster than the lakes. This problem applies particularly to small retention reservoirs, localized at strongly anthropogenic and agricultural areas. The phenomenon is due to natural conditions of water catchment areas and supply of external organic and mineral matter, as well as their low resistance. The degree of natural resistance to the impact of the water catchment area depends on the depth of the reservoir, it's capacity ratio and the length of the coastline. Moreover, it also depends on the coefficient of Schindler (quotient of the total area of the reservoir and catchment area to the total reservoir capacity) and the average annual intensity of water exchange (Chełmicki 2002). The degree of effect of the water catchment on reservoirs as a provider of allochtonous organic and mineral matter can be determined on the basis of total and direct catchment characteristics. The internal production of organic matter resulting from eutrophication, including an phosphorus supply from sediment, may be also a significant reason (Chislock et al. 2013). Resulting bottom sediments have a significant effect on the metabolism of the whole reservoir. In order to take protective action it is necessary to determine the ecological and chemical status of water, assess the impact of catchment management on reservoirs and their vulnerability to degradation process. When the protection and prevention of the reservoir degradation becomes impossible or does not produce the desired results, restoration action should be taken (Lossow 1995). In case of progressive degradation, the most effective method for remediation of waterbodies is considered to be inactivation of phosphorus in sediments.

Organic matter derived from the production within the ecosystem, resulting from the excessive availability of nutrients can have an important contribution to the amount of sediment deposited in reservoirs of stagnant water. Various studies indicate that up to 35% organic matter produced in the euphotic layer of water reservoirs enriches surface layer of bottom sediments (Lehmann et al. 2002, de Junet et al. 2005). The formation of sediment, associated with the mineralization leads in a global scale to the permanent accumulation in the sediment only 0.1 % of net primary production (Lehmann et al. 2002), although paleolimnological studies indicate that the process of accumulation of sediment is directly proportional to the increasing primary production (Tyson 2001). Stored in the bottom sediment material is enriched with significant amounts of phosphorus, which in case of oxygen deficits may harder release to the water as an additional source of this

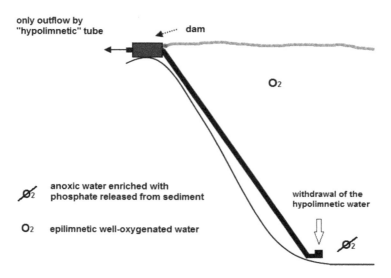

only outflow by
"hypolimnetic" tube

dam

O_2

$\varnothing_2$ anoxic water enriched with
phosphate released from sediment

withdrawal of the
hypolimnetic water

O_2 epilimnetic well-oxygenated water

$\varnothing_2$

Figure 1.   Removing of hypolimnion from lake Kortowskie – Olszewski-tube (Dunalska et al. 2007).

element for plant organisms. Not all forms of phosphorus are released equally from sediments and contribute to an increase in trophic. The sustainability of retention of phosphorus in bottom sediments is determined by the chemical nature of the connection in which this element occurs in deposits. These connections have different solubility. The most mobile is phosphorus contained in organic matter (as a result of mineralization), followed by acting in combination with iron and manganese (Bartoszek et al. 2009). The mobility of phosphorus present in the joints of aluminum is lesser. Phosphorus occurring in combination with calcium has a slight part in the exchange processes between sediment and water (Rzepecki 1997, Kentzer 2001, Golterman 2005).

2   THE LAKE AND RESERVOIR RESTORATION METHODS

Restoration treatments of ecosystems of stagnant water are used in Poland since the 50's. These treatments are based on increasing of exports of biogenic elements from the reservoir, restraining in the bottom sediment or the hindering and slowing their incorporation in the biomass of algae (Lampert & Sommer 2001). In Poland and around the world there were taken many attempts of reclamation of water reservoirs, with varying degree of success. They were conducted by various technical, biological and chemical methods. To technical methods, aimed at reducing the amount of nutrients, was included, for example, the replacement of water in the reservoir or replacement and artificial aeration of the hypolimnion (Fig. 1) (Dunalska et al. 2007).

The method has its limitations: it can be used in reservoirs with a significant flow of water exchange during the year and thermally stratified. This method should not be used in polymictic lakes and when there is another body of water in the river outflow from the reservoir (Lossow & Gawrońska 2000, Zalewski & Izydorczyk 2007, Kostecki 2009). Dredging (removal) of accumulated bottom sediment with phosphorus is very effective method of restoration of shallow, heavily degraded water reservoirs, but far more expensive than others, additionally creating technical problems (Fig. 2). Together with the surface layer of sediment partially eliminated are not only the phosphorus compounds, but also other anthropogenic contaminants, such as heavy metals, pesticides, PAH sediment adsorbed to particles. In addition, dredging is also deepening of silted, shallow body of reservoir. According to Wiśniewski (2004), the removal of a shallow lake sediments leads to the elimination of formed in the reservoir for many years collective of chemical and biological factors, which are essential for the proper functioning of the entire aquatic ecosystem.

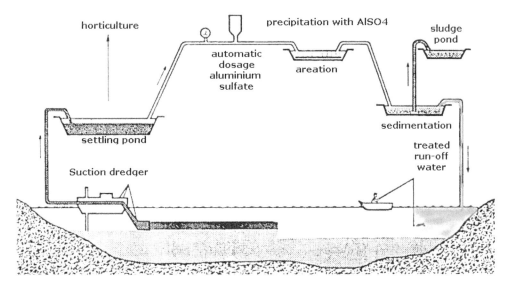

Figure 2. Dredging of the sediment matter from the shallow body of water (Björk 2013).

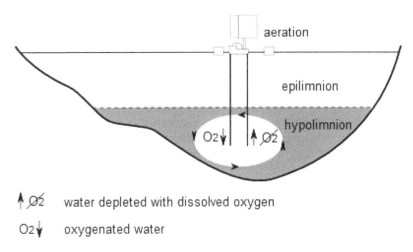

Figure 3. Aeration of the hypolimnion (Podsiadłowski 2013).

Other methods of restoration, such as aeration of water in the reservoir, or better, but much more expensive, oxygenation with pure oxygen, can be treated as an adjuvant treatment, supporting dredging, because when they are used individually, they too often fail (Fig. 3) (Wiśniewski 1999, Jańczak & Kowalik 2004). Often used biomanipulation treatments rely on partial or complete release from the pressure of zooplankton from fish, by reducing of their stocking or even its elimination and thereby on strengthening the pressure of zooplankton on phytoplankton. Improvement obtained in this way is a short-term one (Lampert and Sommer 2001).

Currently, the most effective method of restoration is the inactivation of phosphorus in the bottom sediments (Marcinkowski 2006). Depending on the way of the application of the precipitant, this method leads to the removal of phosphates from the water column (precipitation) and the inhibition of release from bottom sediments (inactivation).

3 METHODS OF CHEMICAL INACTIVATION OF PHOSPHATE IN STAGNANT WATER

The inactivation of phosphorus in bottom sediments up to now was used in dozens of reservoirs in the world. Principally iron, aluminum and calcium compounds, that control the flow of phosphorus in water reservoirs under natural conditions, are used for this purpose (Lossow & Gawrońska 2000, Mastyński & Sawczyn 2005).

Iron compounds (e.g. $Fe_2(SO_4)_3$, $FeCl_2$, $FeCl_3$, PIX), due to the variable valence of this element and the low durability of phosphorus – iron connections under anaerobic conditions, are now rarely used. Effects of inactivation of phosphorus in sediments using iron compounds, often turn out to be short-lived, because the anaerobic conditions in the water above the bottom occur in case of most strongly eutrophic water reservoirs (Perkins & Underwood 2001, Wiśniewski 2003). Inactivation of phosphorus using coagulant PIX-112 (sulphate (VI) iron (III)) with the concomitant oxygenation of water was conducted in the Maltański reservoir (except biomanipulation) and the Trzesiecko lake (Kozak et al. 2007).

Aluminum compounds ($Al_2(SO_4)_3$, PAX, $NaAlO_2$) are characterized by higher efficiency, due to the greater stability of phosphorus – aluminum connection under anaerobic conditions. The alumina in a insoluble form exists in water pH from 6 to 8. The amount of forms dissolved in water, more toxic and dangerous to hydrobionts, increases both at lower and higher pH. At pH above 8.5–9.0 (very common during the phytoplankton bloom) aluminum compounds in the sediment decompose and toxic aluminum ion can be released into water column. The use of aluminum salts to restoration requires particularly careful dosage in case of low alkalinity waters. Especially sulfate (VI) of aluminum, added to water, lowers the pH of water, what leads to increasing of the amount of alumina dissolved in water in toxic forms (Lossow & Gawrońska 2000, Van Hullebusch et al. 2003, Wiśniewski 2004, Haggard et al. 2005). The inactivation of phosphorus using aluminum compounds was carried out in lakes: the Starodworskie lake (sulphate (VI) of aluminum), the Długie lake in Olsztyn and the Głęboczek lake in Tuchola, as well as the Machovo and the Dubice lakes (coagulant PAX 18 – aqueous solution of polyaluminium chloride) (Lossow & Gawrońska 2000, Gawrońska et al. 2003, Gawrońska et al. 2004, Lossow et al. 2004, Mastyński & Sawczyn 2005, Tandyrak 2005, Jankowski 2007). According to Berkowitz et al. (2006) after about 6 months from the use, the aluminum coagulant is aging and the efficiency of phosphate binding is significantly reduced, and the aluminum ion can be released even at neutral pH of water. The connection of coagulants PIX-113 (liquid sulphate (VI) iron (III)) and PAX-18 during restoration of the Wolsztyn lake, reservoirs: the Gołuchów and the Środa lakes, was used in order to increase the efficiency of binding of phosphorus (Małecki 2005, Mastyński & Sawczyn 2005, Gawrońska et al. 2007).

Recently a new, modified zeolite Z2G1 was tested to inactivate phosphorus in the cores of sediment and directly on sediments of the Okaro lake in New Zealand. Experimental studies have shown that a thin layer of Z2G1 covering sediment ("capping") completely inhibits the secretion of phosphate in the aerobic and anoxic conditions. There was no separation of metal from zeolite, or from deposits observed (Gibbs et al. 2011, Özkundakci et al. 2011, Gibbs & Özkundakci 2011). Insensitive to changes in the pH and the redox the coagulant Phoslock®, which is a mixture of sodium bentonite and lanthanum, was used in 2006 to inactivate of phosphate in highly degraded the Klasztorne Małe lake. The disadvantage of this coagulant is a very high price and the vague and imprecise information about the proper dosage. Due to the high cost of treatment, the application of Phoslock® was made only the ferric coagulant PIX (Helman-Grubba 2006, Marcinkowski & Kobos 2006).

Co-precipitation of phosphorus from $Ca(OH)_2$ and $CaCO_3$ was used in the lakes of Canada. According to Prepas et al. (2001b), a lime was less effective than aluminum and iron compounds in inhibiting the release of phosphate from bottom sediments. They observed however, that the application of lime resulted in a marked reduction in phytoplankton biomass in lakes under restoration, with maintaining the pH of water in the natural level of less than 10. To reduce the costs of inactivation in other lakes the combination of expensive $Ca(OH)_2$ and cheaper $CaCO_3$ was used.

It was found that the direct effect of Ca(OH)$_2$ appears to be more effective in removing phosphorus and the reduction of biomass than when used in combination with CaCO$_3$. The maximum reaction observed in the water column of both lakes was 9.3 and 9.1 pH. The Canadian researchers believe that the use of lime to remove phosphorus from the water and inactivation in bottom sediments should be preferred over other factors, as lime is nontoxic and its use is economically justified (Prepas et al. 2001a, 2001b; Zhang et al. 2001). Calcium hydroxide and sodium aluminate (NaAlO$_2$) was fed alternately to eutrophic hypolimnion of the German lake (characterized by high hardness of water), together with simultaneous mixing and aeration of deeper water layers (Koschel et al. 2006). Liming was used in the Polish lakes due to acidification of waters. The sediments of the Flosek Lake after 20 years of liming showed in laboratory studies only a slight tendency to release phosphate. The release proved to be much lower than in other harmonics, non-humic lakes (Rzepecki 1997). Compounds of the calcium do not cause disintegration of cells of algae, what prevents releasing of toxins and nutrients into water. A side effect of their use may be impact on the reaction of water and the disruption of the natural bicarbonate – carbonate balance (Pliński 2009).

Gypsum (sulphate dihydrate (VI) calcium – CaSO$_4\cdot$2H$_2$O) was not previously used in Poland to inactivate of the phosphorus in bottom sediments. It was used experimentally in the hypertrophic Finnish Enäjärvi and Laikkalammi lakes. In the first lake was bottom sediments was covered with an even layer of gypsum deposits in the bases isolated from each other and from the rest of the lake. In the second lake the industrial gypsum (incl. 4% Fe), in dry powder form, was distributed at the surface of water (approximately 500 g of gypsum per m^3 of water) with the help of compressed air. It was found, that gypsum increases the retention capacity of sediments by creating apatite connections and stabilizes them by accumulating in a thin layer of sediment on the surface, preventing the re-suspension process and the release of secondary gases (Salonen & Varjo 2000, Salonen et al. 2001, Varjo et al. 2003).

4 LAKES AND RESERVOIR RESTORATION IN POLAND

During last sixty years the restoration of ca. 50 Polish lakes and reservoirs was conducted. Inactivation of sediment phosphorus was used in 11 reservoirs (PIX and PAX, PAX or PIX and Phoslock®) and mixed methods were used in 15 reservoirs (Jankowski 2007). Applying of chemical preparations is repeated for several consecutive years, although this inactivation is considered one of the cheapest methods of restoration of water reservoirs. This method is particularly recommended for shallow reservoirs. The ratio of the volume of sediment to the water column is significantly larger in shallow reservoirs than in deep reservoirs. The influence of sediments on the metabolism of the whole reservoir is therefore greater. Therefore, the shallow water reservoirs are degraded much faster than the deep ones.

Most of the coagulants used for inactivation does not provide the positive effect of the restoration for a long time, and combining of these formulas does not always produce the expected result. It is not enough to restore the state of the water to the situation prior to degradation. Many feedbacks in the reservoir preserves the state after the changes, and they exhibit resistance to the restoration treatments (Wiśniewski 2004). A chance for the success of restoration of the water can be a comprehensive restoration program. Such program was developed and implemented for the Jelonek and Winiary lakes near Gniezno in 2009–2010. The inactivation of phosphorus in sediments using PIX 111 coagulant (FeCl$_3$) and the Phoslock, applied directly to the sediments together with their simultaneous aeration was the leading method. As a complementary method the biomanipulation was applied. The re-introduction of macrophytes in order to stabilize the water quality in reservoirs was also carried out, as well as the walls of barley straw (secreting natural algistatics) as a supportive treatment was applied to (Trzcińska 2010, Wiśniewski et al. 2010). The results of long-term comprehensive studies of water and bottom sediments, conducted for these lakes, constituted the basis for the development of the program.

5 SUMMARY

Important task is to develop efficient, secure and affordable methods for restoration of such water bodies. This is even more important, that most of these waters also acts as a recreational and economic areas. The natural processes in aquatic ecosystems are the matter in which in the first place we are looking for ways to increase the resilience of the aquatic environment on anthropogenic pollutants introduced by man. Due to the growing demand for water, the most important goal of restoration should be to improve such quality of water in reservoirs that could be used as a source of drinking water, food production and sanitation purposes. Presented methods of bodies of water restorations could be widely used in many cases when the low water standard and eutrophication were noted.

REFERENCES

Bartoszek, L., Tomaszek, J.A., Sutyła, M. 2009. Vertical phosphorus distribution in the bottom sediments of the Solina–Myczkowce Reservoirs. *Environment Protection Engineering* 35(4): 21–29.

Berkowitz, J., Anderson, M.A., Amrhein, Ch. 2006. Influence of aging on phosphorus sorption to alum floc in lake water. *Water Research* 40: 911–916.

Björk, S. 2013. Home page. Retrieved from http://www.vesan.se/3Bjork/2bj_trum.htm

Chislock, M.F., Doster, E., Zitomer, R.A. & Wilson, A.E. 2013. Eutrophication: Causes, Consequences, and Controls in Aquatic Ecosystems. *Nature Education Knowledge* 4(4):10

Chełmicki, W. 2002. Woda. Zasoby, degradacja, ochrona. *PWN*. Warszawa.

De Junet, A., Abril, G., Guérin, F., Billy, I., De Wit, R. 2005. Sources and transfers of particulate organic matter in a tropical reservoir (Petit Saut, French Guiana), a multitracers analysis using δ13C, C/N ratio and pigments. *Biogeosciences Discussions* 2: 1159–1196.

Dunalska, J.A., Wiśniewski, G. & Mientki, Cz. 2007. Assessment of multi-year (1956–2003) hypolimnetic withdrawal from Lake Kortowskie, Poland. *Lake and Reservoir Management* 23: 377–387.

Gawrońska, H., Lossow, K., Łopata, M. 2003. Effectiveness of the aluminium coagulant (PAX) in reducing internal loading in Lake Głęboczek. *Limnological Review* 3: 65–72.

Gawrońska, H., Łossow, K., Grochowska, J., Brzozowska, R. 2004. Rekultywacja jeziora Długiego w Olsztynie metodą inaktywacji fosforu. *V Konferencja Naukowo-Techniczna: Ochrona I Rekultywacja Jezior*. Grudziądz, 12-15.05.2004: 25–32.

Gawrońska, H., Łossow, K., Łopata, M., Brzozowska, R., Jaworska, B. 2007. Rekultywacja jeziora Wolsztyńskiego metodą inaktywacji fosforu. *VI Konferencja Naukowo-Techniczna: Ochrona i Rekultywacja Jezior*. Toruń, 14–16.06.2007: 53–63.

Gibbs, M. & Özkundakci, D. 2011. Effects of a modified zeolite on P and N processes and fluxes across the lake sediment-water interface using core incubations. *Hydrobiologia* 661(1): 21–35.

Gibbs, M., Hickey, C., Özkundakci, D. 2011. Sustainability assessment and comparison of efficacy of four P-inactivation agents for managing internal phosphorus loads in lakes: sediment incubations. *Hydrobiologia* 658(1): 253–275.

Golterman, H.L. 2005. The Chemistry of Phosphate and Nitrogen Compounds in Sediments. *Kluwer Academic Publisher*.

Haggard, B.E., Moore, P.A., DeLaune, P.B. 2005. Phosphorus flux from bottom sediments in Lake Eucha, Oklahoma. *Journal of Environmental Quality* 34: 724–728.

Helman-Grubba, M. 2006. Rekultywacja zdegradowanych akwenów przez inaktywację fosforanów za pomocą preparatu zawierającego lantan. *Przegląd Komunalny* 9(180): 70–71.

Inspekcja Ochrony Środowiska. 2010. Przewodniki metodyczne do badań terenowych i analiz laboratoryjnych elementów biologicznych wód przejściowych i przybrzeżnych. *Biblioteka Monitoringu Środowiska*. Warszawa.

Jankowski, J. 2007. Stan prac rekultywacyjnych w Polsce. *VI Konferencja Naukowo-Techniczna: Ochrona i Rekultywacja Jezior. Toruń, 14–16.06.2007:* 83–94.

Jańczak, J. & Kowalik, A. 2004. Rezultaty stosowania ograniczonej rekultywacji jeziora Jelonek w Gnieźnie. *V Konferencja Naukowo-Techniczna: Ochrona i Rekultywacja Jezior. Grudziądz, 12-15.05.2004:* 55–63.

Kentzer, A. 2001. Fosfor i jego biologicznie dostępne frakcje w osadach jezior różnej trofii. *Rozprawa habilitacyjna*. Wydawnictwo UMK. Toruń.

Koschel, R., Casper, P., Gonsiorczyk, T., Roßberg, R., Wauer, G. 2006. Hypolimnetic Al and $CaCO_3$ treatments and aeration for restoration of a stratified eutrophic hardwater lake in Germany, *Verh. Internat. Verein. Limnol.* 29: 2165–2171.

Kostecki M. 2009. Zmiany wybranych wskaźników jakości wody antropogenicznego zbiornika Pławniowice w pierwszych pięciu latach rekultywacji metodą usuwania wód hipolimnionu. In: J. Ozonek & M. Pawłowska (eds), *Polska Inżynieria Środowiska pięć lat po wstąpieniu do Unii Europejskiej*: 113–127. Lublin: Monografie Komitetu Inżynierii Środowiska PAN.

Kozak, A., Gołdyn, R., Kowalczewska-Madura, K., Dondajewska, R., Podsiadłowski, S. 2007. Rekultywacja zbiornika maltańskiego w Poznaniu, *VI Konferencja Naukowo-Techniczna: Ochrona i Rekultywacja Jezior. Toruń, 14–16.06.2007*: 53–63.

Lampert, W. & Sommer, U. 2001. Ekologia Wód Śródlądowych. *PWN*. Warszawa.

Lehmann, M.F., Bernasconi, S.M., Barbieri, A., Mckenzie, J.A. 2002. Preservation of organic matter and alteration of its carbon and nitrogen isotope composition during simulated and in situ early sedimentary diagenesis. *Geochimica et Cosmochimica Acta* 66(20): 3573–3584.

Lossow, K. 1995. Odnowa jezior. *Ekoprofit*. 5: 11–15.

Lossow, K., Gawrońska, H., Łopata, M., Jaworska, B. 2004. Efektywność rekultywacji polimiktycznego jeziora Głęboczek w Tucholi metodą inaktywacji fosforu. V *Konferencja Naukowo-Techniczna: Ochrona i Rekultywacja Jezior. Grudziądz, 12-15.05.2004*: 131–139.

Lossow, K. & Gawrońska, H. 2000. Przegląd metod rekultywacji jezior. *Przegląd Komunalny* 9(108): 94–106.

Małecki, Z. 2005. Ochrona wód. *EkoTechnika* 4 (36): 8–10.

Marcinkowski, M. & Kobos, J. 2006. Wpływ inaktywacji fosforanów w wodzie i osadach dennych zbiorników eutroficznych przy pomocy modyfikowanej gliny bentonitowej na masowe występowanie cyjanobakterii i zmianę parametrów fizykochemicznych. *Ekol-Unicon*. Retrieved from http://www.ekol-unicon.com.pl/pl/poradnik/publikacje.

Marcinkowski, M. 2006. Rekultywacja jezior a fosfor: wytrącanie z toni wodnej czy inaktywacja w osadach dennych. *Ekol-Unicon*. Retrieved from http://www.ekolunicon.com.pl/pl/poradnik/publikacje.

Mastyński, J. & Sawczyn, P. 2005. Koncepcja Rekultywacji jezior Wielkopolskiego Parku Narodowego na podstawie efektów zastosowania preparatów PAX i PIX w zbiorniku zaporowym w Gołuchowie, Środzie Wlkp. oraz na jeziorach czeskich – Machovo i Dubice. *Seminarium „Nowości w zastosowaniach chemii na oczyszczalniach ścieków – doświadczenia polskie, niemieckie i czeskie". Praga, 6 – 8.09.2005.*

Özkundakci, D., Hamilton, D., Gibbs, M. 2011. Hypolimnetic phosphorus and nitrogen dynamics in a small, eutrophic lake with a seasonally anoxic hypolimnion. *Hydrobiologia* 661(1), 5–20.

Perkins, R.G. & Underwood, G.J.C. 2001. The potential for phosphorus release across the sediment-water interface in an eutrophic reservoir dosed with ferric sulphate. *Wat. Res.* 35 (6): 1399–1406.

Pliński, M. 2009. Przyczyny i skutki zakwitów sinicowych. *IV Ogólnopolskie Warsztaty Sinicowe. Uniwersytet Gdański, Instytut Oceanografii, Regionalne Centrum Sinicowe oraz Polskie Towarzystwo Hydrobiologiczne, Gdynia, 24 czerwca 2009.*

Podsiadłowski, S. 2013. Homa webpage. Retrieved from http://www.aerator.pl.

Prepas, E.E., Babin, J., Murphy, T.P., Chambers, P.A., Sandland, G.J., Ghadouanis, A., Serediak, M. 2001a. Long-term effects of successive $Ca(OH)_2$ and $CaCO_3$ treatments on the water quality of two eutrophic hardwater lakes. *Freshwater Biology* 46: 1089–1103.

Prepas, E.E., Pinel – Alloul, B., Chambers, P.A., Murphy, T.P., Reedyk, S., Sandland, G., Serediak, M. 2001b. Lime treatment and its effects on the chemistry and biota of hardwater eutrophic lakes. *Freshwater Biology* 46: 1049–1060.

Rzepecki, M. 1997. Bottom sediments in a humic lake with artificially increased calcium content: sink or source for phosphorus?. *Water, Air and Soil Pollution* 99 (1/4): 457–464.

Salonen, V-P., Varjo, E., Rantala, P. 2001. Gypsum treatment in managing the internal phosphorus load from sapropelic sediments; experiments on Lake Laikkalammi, Finland. *Boreal Environ. Res.* 6: 119–129.

Salonen, V-P. & Varjo, E. 2000. Gypsum treatment as a restoration method for sediments of eutrophied lakes – experiments from southern Finland. *Environmental Geology* 39 (3–4): 353–359.

Tandyrak, R. 2005. Chemism of bottom sediments from a lake treated with various restoration techniques. *Electronic Journal of Polish Agricultural Universities, Environmental Development* 8 (4): 1–9.

Trzcińska, J. 2010. Doświadczenia samorządu w realizacji projektu — Rekultywacja jezior Jelonek i Winiary w Gnieźnie metodą inaktywacji fosforu w osadach dennych dofinansowanego z Programu Life +. *VII Konferencja Naukowo-Techniczna: Ochrona i rekultywacja jezior. Toruń, 07–09.10.2010*: 153–156.

Tyson, R.V. 2001. Sedimentation rate, dilution, preservation and total organic carbon: some results of a modelling study. *Organic Geochemistry* 32 (2): 333–339

Van Hullebusch, E., Auvray, F., Deluchat, V., Chazal, Ph.M., Baudu, M. 2003. Phosphorus fractionation end short-term mobility in the surface sediment of a polymictic shallow lake treated with a low dose of alum (Courtille Lake, France). *Water, Air, and Soil Pollution* 146: 75–91.

Varjo, E., Liikanen, A., Salonen, V.-P., Martikainen, P.J. 2003. A new gypsum-based technique to reduce methane and phosphorus release from sediments of eutrophied lakes: (Gypsum treatment to reduce internal loading). *Water Research* 37: 1–10.

Wiśniewski, R., Ślusarczyk, J., Kaliszewski, T., Szulczewski, A., Nowacki, P. 2010a. „Proteus", a new device for application of coagulants directly to sediment during its controlled resuspension. *Verh. Internat. Vorein. Limnol.* 30(9): 1421–1424.

Wiśniewski, R. 2003. Fosforany w osadach hypertroficznych jezior. Usuwać, czy immobilizować?. *III Ogólnopolska Konferencja Naukowo-Techniczna: Postęp w Inżynierii Środowiska. Rzeszów-Polańczyk, 25–27.09.2003*: 265–274.

Wiśniewski, R. 2004. Phosphates in sediments of hypertrophic Łasińskie lake. Remove or immobilize?. *Environment Protection Engineering* 30(4): 161–169.

Wiśniewski, R. 1999. Próby inaktywacji fosforanów w osadach dennych i zahamowania zakwitu sinic w jeziorze Łasińskim jako potencjalne metody rekultywacji. *I Ogólnopolska Konferencja Naukowo-Techniczna: Postęp w Inżynierii Środowiska. Rzeszów-Polańczyk, 30.09-2.10.1999*: 189–202.

Zalewski, M. & Izydorczyk, K. 2007. Ekohydrologia – systemowe podejście do ochrony i rekultywacji jezior. *VI Konferencja Naukowo-Techniczna: Ochrona i Rekultywacja Jezior. Toruń, 14–16.06.2007*: 211–219.

Zhang, Y., Ghadouani, A., Prepas, E.E., Pinel-Alloul, B., Reedyk, S., Chambers, P.A., Robarts, R.D., Methot, G., Raik, A., Holst, M. 2001. Response of plankton communities to whole-lake $Ca(OH)_2$ and $CaCO_3$ additions in eutrophic hardwater lakes. *Freshwater Biology* 46: 1105–1119.

Progress in Environmental Engineering – Tomaszek & Koszelnik (eds)
© 2015 Taylor & Francis Group, London, ISBN: 978-1-138-02799-2

The use of keramsite grains as a support material for the biofilm in moving bed technology

A. Masłoń & J.A. Tomaszek

Department of Chemistry and Environmental Engineering, Faculty of Civil and
Environmental Engineering, Rzeszów University of Technology, Rzeszów, Poland

ABSTRACT: The paper presents studies on wastewater treatment in a sequencing batch reactor using keramsite grains as the moving bed biofilm PCMBSBBR (Porous Carrier in Moving Bed Sequencing Batch Biofilm Reactor). With municipal wastewater and PCMBSBBR technology, the study demonstrated high average removal efficiencies – of 96.5; 85.2; and 91.3% respectively for COD, TN and TP. Analyzed technology has examined the effects of wastewater treatment systems with respect to MBSBBR with conventional plastic carriers. Unfortunately, this type of keramsite grains that was used in the study proved to be unsuitable in moving bed biofilm technology. Analyzed keramsite had relatively low mechanical strength, which appears during the study. Fur-ther studies using keramsite grains are required. It is necessary to use of keramsite with better mechanical properties than ever before to grain and clashed not crumble. Key conclusions of this studies have identified a new line of research into the use of powdered form of keramsite in activated sludge technology in sequencing batch reactor.

1 INTRODUCTION

Evolution of wastewater treatment technologies in biofilm systems is actually observed. Biofilm systems can be separated into categories in many ways, one of these is as follows: technologies using suspended or fluidized biofilm carriers (e.g. Moving Bed Biofilm Reactor (MBBR), biologically aerated filter, granular technologies, fluidized bed bioreactors) and technologies using fixed biofilm carrier (e.g. trickling filter, rotating biological contactor) (Nicolella et al. 2000, Szilágyi et al. 2011). More modern wastewater treatment technologies in biofilm systems are developed and tested (Gullicks et al. 2011, Szilágyi et al. 2011). Various types of biofilm reactors are used or designed in laboratory scale (Krzemieniewski & Rodziewicz 2005, Zieliński & Krzemieniewski 2007, Szilágyi et al. 2013). New unconventional biofilm support and carriers are sought and tested (May-Esquivel et al. 2008, Szilágyi et al. 2011, Makkulath & Thampi 2012, Darvishi Cheshmeh Soltani et al. 2013, Jurecska et al. 2013, Spychała et al. 2013).

The moving bed biofilm technology is still especially developed technology of wastewater treat-ment. This technology is based on the biofilm principle, which implies that the microorganisms treating the wastewater are grown on the surface of a carrier material in the treatment process. The moving bed biofilm technology is move freely elements in the reactor having a density close to the density of water and the large specific surface area, which are base for the growth of the microorganisms. In recent years researches focusing on hybrid systems combining the advantages of suspended growth and biofilm systems have increased. Moving bed biofilm reactors holding carrier elements freely moving in the reactor have been developed as one of the most attractive hybrid systems (Ødegaard et al. 1994). In the last 20 years important study has been conducted on pilot or full-scale MBBR systems for the removal of organic carbon, nitrogen and phosphorus from wastewaters (Ødegaard et al. 1994, Pastorelli et al. 1997, Andreottola et al. 2000, Ødegaard et al. 2000, Rusten et al. 2006). The idea of moving bed biofilm technology is to use the advantages

Table 1. Characteristics of various types of plastic biomass carriers used in the moving bed biofilm reactors (Podedworna & Żubrowska-Sudoł 2006, Podedworna et al. 2009).

Name	Material	Form	Size carrier [mm]	Specific surface area $[m^2/m^3_{carrier}]$
Linpor	PU*	cube	$14 \times 14 \times 14$	1000
Kaldnes K1	PE**	cylinder	Ø10, H 7	500
Kaldnes K3	PE**		Ø25, H 12	500
Natrix C2	PE**	truncated cone	Ø36, H 30	220
Natrix M2	PE**		Ø64, H 50	200
EvU®-Perl	PVA***	cylinder	Ø5, H 8	800
Biolox	PE**	cylinder	Ø14, H 8	640
Bioflow 9	PP****	cylinder	Ø9, H 7	800
FLOCOR RMP	PP****	cylinder	Ø15–20, H 20–30	160
Fleece 2	PE**	chopped fibers	$10 \times 10 \times 2$	500
Newfloat	PE**	irregular spherical shapes	20	700

*PU – polyurethane foams
**PE – polyethylene
***PVA – recycled polyvinyl alcohol
****PP – polypropylene

of both activated sludge and biofilter to the exclusion of their defects (Podedworna & Żubrowska Sudoł 2006). Several different types of biomass carriers are used in MBBR technology. This are mainly synthetic materials from plastic. Table 1 presents the characteristics of the sample biofilm carriers used in MBBR reactors.

Drawing on the fact that SBR reactors in some cases display advantages over continuous-flow systems and biofilm processes over activated sludge technology, there is considered to be a possibility of improving nutrient the process by which nutrients are removed from wastewater in SBR systems, through the additional use of biofilm biomass (Masłoń & Tomaszek 2009b). One solution of this type is the 'so-called' MBSBBR (*Moving Bed Sequencing Batch Biofilm Reactor*), in which biofilm carriers are elements swimming freely within the whole reactor volume (Pastorelli et al. 1999, Helness & Ødegaard 1999, Helness & Ødegaard 2001, Żubrowska-Sudoł 2004, Masłoń & Tomaszek 2008, Podedworna et al. 2009). The bed biofilms in MBSBBR systems are made from different materials (e.g. plastics, polyurethane foam). They are characterized by different form and diversified specific surface area ($800–1000 \, m^2 \cdot m^{-3}$), this offering better conditions for microorganisms to develop. As the form of carrier and bed type together determine the operation of biological wastewater treatment (Żubrowska-Sudoł 2004), it is advisable to seek out new, highly effective carriers. It is possible to use highly-porous materials as biofilm carriers in MBSBBR reactors (Shin & Park 1991, Masłoń & Tomaszek 2009a, 2009b, 2009c).

This paper offers an analysis of studies on wastewater treatment in a sequencing batch reactor using keramsite grains as the moving bed biofilm PCMBSBBR (*Porous Carrier in Moving Bed Sequencing Batch Biofilm Reactor*).

2 MATERIALS AND METHODS

Studies in a lab-scale model MBSBBR were carried out (Fig. 1). The keramsite grains (kind of light expanded clay aggregate) were used as a support material for the biofilm. The filling rate was about 10% of working volume. Table 2 denotes the characteristics of the keramsite material. Bearing in mind the use of porous material as a biomass carrier, the name for the analysed system – PCMBSBBR (*Porous Carrier in Moving Bed Sequencing Batch Biofilm Reactor*) – was arrived by the authors. The system was operated over 8-hour cycles (Table 3). Dissolved oxygen (DO)

Figure 1.   The laboratory MBSBBR reactor used in this study.

Table 2.   Characteristics of the porous biomass carrier used in this study and the technological parameters of the PCMBSBBR system.

Parameter, unit		
Support material	keramsite	
Size carrier	mm	4–8
Specific surface area	$m^{-2} \cdot m^{-3}$	900
Specific gravity	$g \cdot dm^{-3}$	0.85
Total reactor volume	dm^3	18.5
Working reactor volume	dm^3	15.0
Total volume of support material	dm^3	1.5
Mixed liquor suspended solids (MLSS)*	$g \cdot dm^{-3}$	3.68–5.21
Volumetric exchange ratio (VER)	–	0.4
Hydraulic retention time (HRT)	h	20.0
Sludge retention time (SRT)	d	10.0
Organic loading rate (OLR)	$g \, COD \cdot dm^{-3} \cdot d^{-1}$	0.432–0.972

concentration was maintained at a level of 2–3 mg $O_2 \cdot dm^{-3}$. A volume of 6.0 dm^3 of wastewater was supplied during the cycle. It was active sludge from the Rzeszów WWTP that was used as the inoculum of the PCMBSBBR system. The SRT was established as 10 days with regard to different requirements as regards SRT in the cases of the nitrification, denitrification and biological dephosphatation processes. The SRT control entailed the discharge of wastewater and activated sludge at the end of the reaction phase. The study was carried out using a synthetic wastewater consisting of pepton, ammonium acetate and glycogen, with mineral salts and phosphorus compounds (Table 4) (Masłoń & Tomaszek 2009a).

Table 3. Distribution of the operational cycles of the PCMBSBBR reactor.

Phase									
Filling									
Reaction	Mixing								
	Aeration								
Settling									
Decantation									
Idle									
Time [h]		0.5	0.5	2.0	1.0	2.0	1.25	0.5	0.25

Table 4. The characteristics of the raw wastewater used in this study.

Parameter, unit	Min.	Max	Avg	SD	Mediana
COD, mg $O_2 \cdot dm^3$	360.0	810.0	680.5	96.05	665.0
N-NH$_4$, mg N $\cdot dm^{-3}$	30.8	49.6	41.3	2.96	41.2
TKN, mg N $\cdot dm^{-3}$	65.0	86.9	80.9	4.42	83.1
TN, mg N $\cdot dm^{-3}$	66.4	88.0	81.6	4.71	83.0
P-PO$_4$, mg P $\cdot dm^{-3}$	7.5	15.0	10.7	1.74	10.7
TP, mg P $\cdot dm^{-3}$	13.5	23.75	17.7	1.35	17.6
COD/TP ratio	20.0	47.4	38.9	5.71	39.7
COD/TN ratio	5.1	9.3	8.3	0.83	8.7

In this study was defined composition of inflowing and treated wastewaters, the physical – chemical and concentration of activated sludge in the system. The researches included the determination of COD, Total Nitrogen (TN), Total Kjeldahl Nitrogen (TKN), Ammonium Nitrogen (N-NH$_4$), Nitrite Nitrogen (N-NO$_2$), Nitrate Nitrogen (N-NO$_3$), Total Phosphorus (TP), Phosphate Phosphorus (P-PO$_4$). The chemical analyses were performed in accordance to the Polish Standard Methods (PN). DO and pH were also routinely monitored using a DO meter and the pH meter.

This paper presents the results of 20 series of measurements during the 10 weeks of system operation (after 6 weeks' adaptation).

3 RESULTS AND DISCUSSION

Obtained results of studies on wastewater treatment in the PCMBSBBR summarizes the characteristics of the proposed system in terms of organic carbon and nutrients removal.

The experiments were conducted at a food-to-microorganism ratio (F/M) of 0.072–0.168 g COD $\cdot$ g MLSS$^{-1} \cdot d^{-1}$ (Masłoń & Tomaszek 2009c, Tomaszek & Masłoń 2010). Fig. 2 shows COD removal performance of the PCMBSBBR system for the entire experimental period. During the testing period the COD decrease reached a level of 94.4–99.2% at the volumetric total COD-loading rate 0.432–0.972 g COD $\cdot$ dm$^{-3} \cdot d^{-1}$. A high degree of removal of organic compounds was achieved corresponding to values in effluent of between 4.8–36.8 mg $O_2 \cdot dm^{-3}$ of COD. The COD removal rate well correlated with COD loading rate (Fig. 3). This indicates that the slope of the linear regression can be considered to the average removal efficiency of COD, of which the system was 96.5% with R^2 of 0.995. The suspended biomass growth amounted to 0.73 g MLSS $\cdot$ g COD^{-1}.

In the PCMBSBBR system it was possible to achieve a high degree of TN and TKN removal from wastewater at the levels 75.9–92.2% and 81.3–97.7%, respectively (Fig. 4). This was possibly a reflection of a high degree of nitrification. During the studies, the concentrations of ammonium

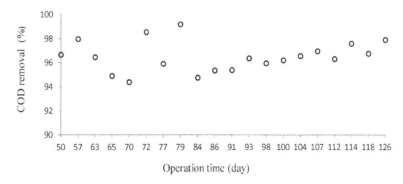

Figure 2.   The efficiency of COD removal in the PCMBSBBR system.

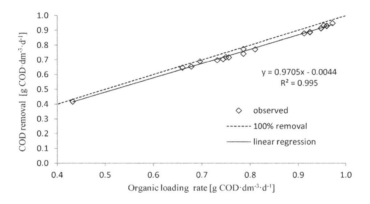

Figure 3.   The performance of COD removal in the PCMBSBBR system.

nitrogen and TKN were observed to be low. The effluent N-NH$_4$ and TKN concentration were of 0.3–6.3 mg N · dm^{-3} and 1.96–3.22 mg N · dm^{-3}. This corresponds to a nitrification efficiency of 79.5 to 99.3%. Significantly higher efficiency of nitrification in MBSBBR system was obtained by Podedworna & Żubrowska-Sudoł (2009). Authors achieved an nitrifiaction between 93.6 and 99.7% (97.9% on average). The volumetric total nitrogen loading rate amounted to 0.08–0.106 g N · dm^{-3} · d^{-1}, whereas the average TN-load of activated sludge was 16 mg N · g MLSS^{-1} · d^{-1}.

Analysis of the PCMBSBBR cycle showed nitrogen removal to result from classical nitrification and denitrification (Masłoń & Tomaszek 2009c, Tomaszek & Masłoń 2010). The dissolved oxygen concentration in the aerobic subphase reaction was maintained in the range 2.0–3.0 mg O$_2$ · dm^{-3}. Concentrations of N-NH$_4^+$ to levels below 1.0 mg were observed in the effluent. However, complete nitrification wasn't achieved in this system. The effluent nitrite and nitrate concentrations were at levels of 0.06–4.70 mg N-NO$_2$ · dm^{-3} and 0.55–5.50 mg N-NO$_3$ · dm^{-3}, respectively. To observe the effect of loading rates and COD/TN ratio on the PCMBSBBR reactor performance, Figs. 5–7 were developed.

The studies didn't show a statistically significant impact of COD/TN ratio on nitrogen removal efficiency (Fig. 7). This shows considerable immunity of reactor with keramsite grains to the variable composition of the wastewater flowing into the system. The effluent TN concentration was between 6.9 and 17.5 mg N · dm^{-3} with an average of 11.9 mg N · dm^{-3}.

In the PCMBSBBR system, effective biological dephosphatation was achieved (Masłoń & Tomaszek 2009a, 2009c). Biological phosphorus removal was based on the release of orto-P in anaerobic phase and an increased uptake in the later aerobic stage. Removal of total phosphorus achieved values of 75.0–97.5% at a TP-load rate of 16–29 mg P · dm^{-3} · d^{-1} (Fig. 8). The average

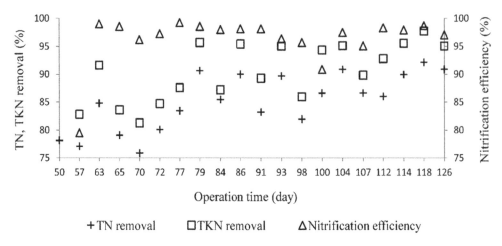

Figure 4.   The efficiency of nitrogen removal in the PCMBSBBR system.

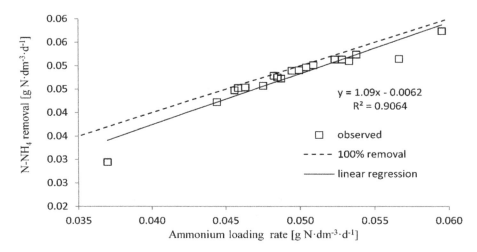

Figure 5.   N-NH₄ removal performance in the system.

activated sludge loading was 2.0–5.0 mg P · g MLSS^{-1} · d^{-1}. The obtained TP removal efficiencies were very variable with average removal rate of 19.4 mg P · dm^{-3} · d^{-1} (Fig. 9).

The TP removal from municipal wastewater is dependent on the COD:TP ratio in influent (Fig. 10). In our experiment the highest efficiency of TP removal was achieved at the COD:TP ratio above 40 (Fig. 11). Considering the high variation of the effluent TP the observed slight increasing tendency of the phosphorus removal with the COD load increase is statistically significant (Fig. 12).

4   CONCLUSION

Unusual properties of keramsite grains cause an increase in the interest for use in the wastewater technology and the activated sludge system. The use of expanded clay in wastewater treatment leads to the use of the specific properties of porous sorption and thermal insulation (Masłoń & Tomaszek 2010).

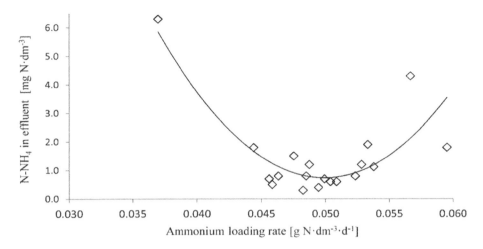

Figure 6.   The impact of ammonium loading rate on the quality effluent from the reactor.

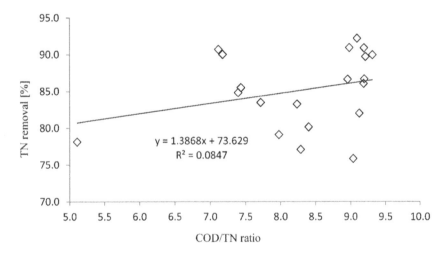

Figure 7.   The impact of influence COD/TN ratio on nitrogen removal.

Applied in a PCMBSBBR, the analyzed keramsite, has a cellular structure of grains, which contributes to the generation of specific aerobic-anaerobic conditions for biofilm microorganism growth. It also has a thermo-insulating characteristic causing stabilization of temperature and accumulation of warmth within the biofilm. The new PCMBSBBR technology demonstrates a good possibility for integrated carbon, nitrogen and phosphorus removal from municipal wastewater to be achieved. The studies conducted under a COD loading rate of $0.432–0.972$ g $COD \cdot dm^{-3} \cdot d^{-1}$ and an SRT of 10 days gave a high average removal efficiency of 96.5; 85.2; and 91.3% for COD, TN and TP respectively. A stable nitrification process (96.3% of ammonium nitrogen removal) can also be obtained. The PCMBSBBR reactor demonstrated a good resistance to load variation of organic matter and nutrients. Analyzed technology has examined the effects of wastewater treatment systems with respect to MBSBBR with conventional plastic carriers. In studies conducted so far maximum removal of organic carbon, nitrogen and phosphorus respectively: 92–96, 81–90 and 99% (Podedworna & Żubrowska-Sudoł 2001), 93, 67 and 88% (Kim et al. 2003), 96, 62 and 84% (Manoj Kumar & Chaudhari 2003), 90–98, 61–87, 91–99% (Podedworan & Żubrowska-Sudoł 2008); 93.5, 82.6, 84.1% (Yang et al. 2010), 87.7–97.6, 65.4–81.0, 95.0–97.2 (Jin et al. 2012).

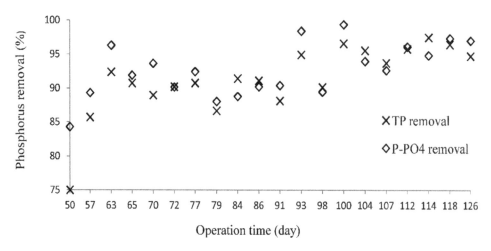

Figure 8. The efficiency of phosphorus removal in the PCMBSBBR system.

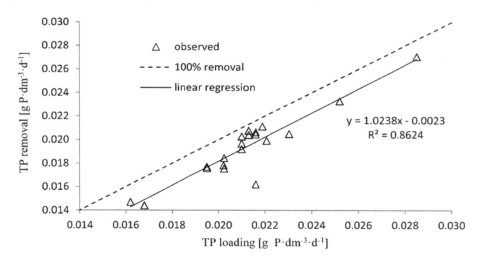

Figure 9. The performance of TP removal in the PCMBSBBR system.

The use of porous carriers due to their specific properties can significantly improve the efficiency of moving bed technology. Researches of moving bed with a high porosity material were conducted using synthetic materials include polyurethane (Shin & Park, 1991; Manoj Kumar & Chaudhari 2003, Jianlong et al. 2000), polyethylene (Cho et al. 2001) and expanded clay (Liapor®) (Valvidia et al. 2007) so far. The first reports of the use of porous materials in a sequencing batch reactor PBCSBR (*Porous Biomass Carrier Sequencing Batch Reactor*) showed a high removal efficiency of organic matter, nitrogen and phosphorus, respectively, over 98% – of COD, 99% – N-NH₄, 95% – TP (Shin & Park 1991).

The use of porous materials as biomass carrier is fully justified, because the high porosity associated with the size of the surface area increases the efficiency of the biofilm. It provides a large surface area of the substrate and improves the diffusion microbiological substrates for the interior (Jianlong et al. 2000, Masłoń & Tomaszek 2009b). Keramsite have a cellular structure of grains, which can help to generate oxygen-specific anaerobic conditions for development of the biofilm microorganisms. There was complete biofilm both open and closed internal pores of keramsite grains (Masłoń & Tomaszek 2009c). Analyzed keramsite had relatively low mechanical

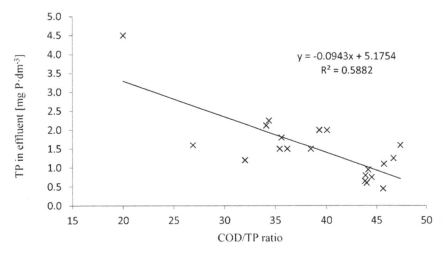

Figure 10.   The impact of influence COD/TP ratio on phosphorus removal.

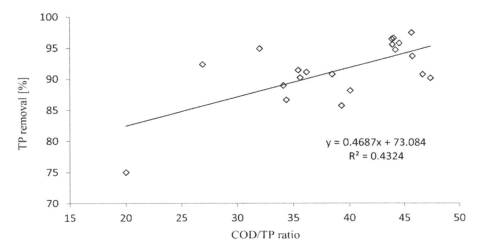

Figure 11.   The impact of influence COD/TP ratio on efficiency of phosphorus removal.

strength, which appears during the study. Keramsite grains underwent abrasion and the resulting microparticles were incorporated into the structure of activated sludge and discharged from the system with excessive sediment. It is believed that the presence of microparticles of keramsite in the activated sludge could favorably influence the processes of biological wastewater treatment as the dosage powdered natural zeolite (Anielak & Piaskowski 2005, Anielak 2006, He et al. 2007), bentonite (Lee et al. 2002, Wiszniowski et al. 2007) and clinoptilolite (Lee et al. 2002). Keramsite microparticles were "dead weight" that increases the density of activated sludge and sedimentation velocity while reducing sludge volume index. Studies have also shown that a small amount of keramsite grains underwent "irreversible sedimentation" as a result of significant impregnability of porous material. This precludes its use as biofilm carriers for the moving bed with such a grain size.

Further studies using keramsite grains are required. It is necessary to use of keramsite with better mechanical properties than ever before to grain and clashed not crumble.

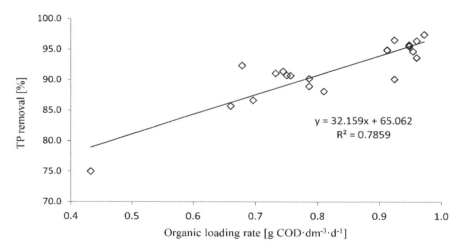

Figure 12.   The impact of OLR on efficiency of phosphorus removal.

Key conclusions of this studies have identified a new line of research into the use of powdered form of keramsite in activated sludge technology in sequencing batch reactor (Masłoń & Tomaszek 2013).

REFERENCES

Agridiotis, V., Forster, FCIWEM C.F., Balavoine, C., Wolter, C., Carliell-Marquet C. 2006. An examination of the surface characteristics of activated sludge in relation to bulking during the treatment of paper mill wastewater. *Water and Environment Journal* 20: 141–149.
Agridiotis, V., Forster, C.F., Carliell-Marquet, C. 2007. Addition of Al and Fe salts during treatment of paper mill effluents to improve activated sludge settlement characteristics. *Bioresource Technology* 98: 2926–2934.
Anielak, A.M. 2006. Niekonwencjonalne metody usuwania substancji biogennych w bioreaktorach sekwen-cyjnych. *Gaz Woda i Technika Sanitarna* 2: 23–27.
Anielak, A.M. & Piaskowski, K. 2005. Influence of zeolites on kinetics and effectiveness of the process of sewage biological purification in Sequencing Batch Reactors. *Environmental Protection Engineering* 2: 31, s. 21–31.
Barnett, J., Richardson, D., Stack K., Lewis T. 2012. Addition of trace metals and vitamins for the optimization of a pulp and paper mill activated sludge wastewater treatment plant. *Appita Journal* 65(3): 237–243.
Bartkiewicz, B. & Umiejewska, K. 2004. Zmiany w technologii oczyszczania ścieków i przeróbki osadów ściekowych. *Instal* 2: 42–46.
Bazeli, M. 2005. Wpływ wybranych koagulantów glinowych i żelazowych na dominujące bakterie nitkowate. *Seminarium Naukowo – Techniczne pt. „Nowości w zastosowaniu chemii na oczyszczalniach ścieków w oparciu o doświadczenia polskie, niemieckie i czeskie". Praga – Teplice. Materiały seminaryjne*: 38–48.
Bazeli, M. 2008. Kryteria selekcji bakterii nitkowatych w oczyszczalniach komunalnych i przemysłowych. *Seminarium Naukowo-Techniczne pt. „Nowe zastosowania chemii w technologii oczyszczania ścieków komunalnych i przemysłowych". Mikołajki, 10–12 września 2008 r. Materiały seminaryjne*: 12–19.
Bazeli, M. 2009. Rola analizy mikroskopowej w walce z bakteriami nitkowatymi. *Seminarium Naukowo-Techniczne pt. "Różne aspekty chemicznych procesów oczyszczania ścieków ze szczególnym uwzględnieniem ich energochłonności". Sopot, 30 września – 1 października 2009 r.*: 73–78.
Bidault, A., Clauss, F., Helaine, D., Balavoine, C. 1997. Floc agglomeration and structuration by a specific talc mineral composition. *Water Science and Technology* 36(4): 57–68.
Böhler, M. & Siegrist, H. 2004. Partial ozonation of activated sludge to reduce excess sludge, improve denitrification and control scumming and bulking. *Water Science and Technology* 49(10): 41–49.
Cantet, J., Paul, E., Clauss, F. 1996. Upgrading performance of an activated sludge process through addition of talqueous powder. *Water Science and Technology* 34(5–6): 75–83.

Chu, L., Yan, S., Xing, X.-H., Sun, X., Jurcik, B. 2009. Progress and perspectives of sludge ozonation as a powerful pretreatment method for minimization of excess sludge production. *Water Research* 43(7): 1811–1822.

Chudoba, P. & Pannier, M. 1994. Use of Powdered Clay to Upgrade Activated Sludge Process. *Environmental Technology* 15: 863–870.

Clauss, F., Balavoine, C., Hélaine, D., Martin, G., (1999). Controlling the settling of activated sludge in pulp and paper wastewater treatment plants. *Water Science and Technology* 40(11–12): 223–229.

Comas, J., Rodríguez-Roda, I., Gernaey, K.V., Rosen, C., Jeppsson, U., Poch, M. 2008. Risk assessment model ling of microbiology-related solids separation problems in activated sludge systems. *Environmental Modelling & Software* 23: 1250–1261.

Czerwionka, K. 2007. Zastosowanie reagentów chemicznych do zwalczania bakterii nitkowatych *Nocardia*. *Seminarium Naukowo-Techniczne pt. „Różnorodność zastosowań chemii w oczyszczaniu ścieków, obróbce osadów i zwalczaniu odorów". Darłówek, 12–14 września 2007 r. Materiały seminaryjne:* 67–79.

Czerwionka, K. 2008. Zwalczanie bakterii nitkowatych typu *Nostocoida limicola I* 0041 w oczyszczalni w Tłuczewie. *Seminarium Naukowo-Techniczne pt. „Nowe zastosowania chemii w technologii oczyszczania ścieków komunalnych i przemysłowych". Mikołajki, 10–12 września 2008 r. Materiały seminaryjne:* 20–27.

Davoli, D., Madoni, P., Guglielmi, L., Pergetti, M., Barilli, S. 2002. Testing the effect of selectors in the control of bulking and foaming in full scale activated – sludge plants. *Water, Science and Technology* 46(1–2): 495–498.

Dobiegała, E. & Remiszewska-Skwarek, A. 2009. Doświadczenia ze zwalczania bakterii nitkowatych w czasie modernizacji oczyszczalni ścieków „Dęgoborze". *Seminarium Naukowo-Techniczne pt. "Różne aspekty chemicznych procesów oczyszczania ścieków ze szczególnym uwzględnieniem ich energochłonności". Sopot, 30 września – 1 października 2009 r.:* 79–90.

Dominowska, M. 2011. Zastosowanie wodnego roztworu polichlorku glinu (PAX) do zwalczania bakterii nitkowatych na Oczyszczalni Ścieków „Pomorzany" w Szczecinie. *Seminarium Naukowo-Techniczne pt. „Chemia w uzdatnianiu wody i oczyszczaniu ścieków. Nowe zastosowania na bazie 20 lat doświadczeń", Szczecin, 4–7 października 2011 r. Materiały seminaryjne:* 135–143.

Drzewicki, A. 2005. Znaczenie morfologii kłaczka w procesie oczyszczania ścieków metodą osadu czynnego. *Gaz, Woda i Technika Sanitarna* 9: 26–27.

Drzewicki, A. 2007. Znaczenie obserwacji mikroskopowych osadu czynnego. *Gaz, Woda i Technika Sanitarna* 11: 22–23.

Drzewicki, A. 2009. Effect of application of polyaluminium chloride on reducing exploitation problems as the wastewater treatment plant in Olsztyn. *Polish Journal of Natural Sciences* 24(3): 158–168.

Drzewicki, A. & Tomczykowska, M. 2008. Wpływ rozwiązań operacyjnych na ograniczenie problemów eksploatacyjnych w skali technicznej wywołanych nadmiernym rozwojem Microthrix Parvicella w osadzie czynnym. *Forum Eksploatatora* 1: 26–29.

Dymaczewski, Z. 2011. *Poradnik eksploatatora oczyszczalni ścieków.* Wyd. PZiTS. Poznań.

Eikelboom, D.H. & Grovenstein, J. 1998. Control of bulking in a full scale plant by addition of talc (PE 8418). *Water Science and Technology* 37(4–5): 297–301.

Eikelboom, D.H. & van Buijsen, H.J.J. 1999. *Podręcznik mikroskopowego badania osadu czynnego.* Wyd. Seidel-Przywecki, Szczecin.

Fiałkowska, E. & Pajdak-Stós, A. 2008. Preliminary studies on the role of *Lecane rotifers* in activated sludge bulking control. *Water Research* 42(10–11): 2483–2490.

Andreottola, G., Foladori P., Ragazzi M. 2000. Upgrading of small wastewater treatment plant in a cold climate region using a moving bed biofilm reactor (MBBR) system. *Water Science and Technology* 41(1): 177–185.

Anielak, A.M., Piaskowski, K. 2005. Influence of zeolities on kinetics and effectiveness of the process of sewage biological purification in sequencing batch reactors. *Environmental Protection Engineering*: 2, 31, 21–31.

Anielak, A.M. 2006. Niekonwencjonalne metody usuwania substancji biogennych w bioreaktorach sekwen-cyjnych (Eng. Unconventional methods of removing biogenic substances in sequence bioreactors). *Gaz, Woda i Technika Sanitarna*: 2, 23–27.

Cho, B-C., Chang, C-N., Liaw, S-L., Huang, P-T. 2001. The feasible sequential control strategy of treating high strenght organic nitrogen wastewater with sequencing batch biofilm reactor. *Water, Science and Technology*: 43(3), 115–122.

Darvishi Cheshmeh Soltani, R., Rezaee, A., Godini, H., Khataee, A.R., Jorfi, S. 2013. Organic Matter Removal Under High Loads in a Fixed-Bed Sequencing Batch Reactor with Peach Pit as Carrier. *Environmental Progress & Sustainable Energy*: 32(3), 681–687.

Gullicks, H., Hasan, H., Das, D., Moretti, C., Hung, Y.T. 2011. Biofilm fixed film systems. *Water:* 3, 843–868.

He, S.B., Xue, G., Kong, H.N., Li, X. 2007. Improving the performance of sequencing batch reactor (SBR) by the addition of zeolite powder. *Journal of Hazardous Materials:* 142(1–2): 493–499.

Helness, H. & Ødegaard, H. 1999. Biological phosphorus removal in a sequencing batch moving bed biofilm reactor. *Water Science and Technology* 40(4–5): 161–168.

Helness, H. & Ødegaard, H. 2001. Biological phosphorus and nitrogen removal in a sequencing batch moving bed biofilm reactor. *Water Science and Technology* 43(1): 233–240.

Jianlong, W., Hanchang, S., Yi, Q. 2000. Wastewater treatment in a hybrid biological reactor (HBR): effect of organic loading rates. *Process Biochemistry:* 36, 297–303.

Jin, Y., Ding, D., Feng, C., Tong, S., Suemura, T., Zhang, F. 2012. Performance of sequencing batch biofilm reactors with different control systems in treating synthetic municipal wastewater. *Bioresource Technology:* 104, 12–18.

Jurecska, L., Barkács, K., Kiss, É., Gyulai, G., Felföldi, T., Tör, B., Kovács, R., Záray, G. 2013. Intensification of wastewater treatment with polymer fiber-based biofilm carriers. *Microchemical Journal:* 107, 108–114.

Kim, H., Rhu, D., Hwang, H., Choi, E. 2003. Performance of a hybrid SBR with fixed bed and suspended growth. *Water Science and Technology* 48(11–12): 309–317.

Krzemieniewski, M., Rodziewicz, J. 2005. Nitrogen compounds removal in a rotating electrobiological contactor. *Environmental Engineering Science:* 22. 6, 816–822.

Lee, H.S., Park, S.J., Yoon, T.I. 2002. Wastewater treatment in a hybrid biological reactor using powdered minerals: effects of organic loading rates on COD removal and nitrification. *Process Biochemistry:* 38, 81–88.

Makkulath, G., Thampi, S.G. 2012. Performance of coir geotextiles as attached media in biofilters for nutrient removal. *International Journal of Environmental Sciences:* 3(2), 784–794.

Manoj Kumar, B., Chaudhari, S. 2003. Evaluation of sequencing batch reactor (SBR) and sequencing batch biofilm reactor (SBBR) for biological nutrient removal from simulated wastewater containing glucose as carbon source. *Water Science and Technology* 48(3): 73–79.

Masłoń, A. & Tomaszek, J.A. 2008. Innowacyjne rozwiązania sekwencyjnych reaktorów porcjowych stosowane w oczyszczaniu ścieków (Eng. Innovative types of sequencing batch reactors used in wastewater treatment). *Inżynieria i Ochrona Środowiska* 11(4): 431–453.

Masłoń, A. & Tomaszek, J.A. 2009a. Preliminary studies on wastewater treatment in a PCMBSBBR reactor. *IWA 2nd Specialized Conference on „Nutrient Management in Wastewater Treatment Processes". Krakow, Poland, 6–9th September 2009. Conference proceedings: 1237–1242.*

Masłoń, A. & Tomaszek, J.A. 2009b. Przegląd literatury nowych rozwiązań technologicznych reaktorów sekwencyjnych z błoną biologiczną (Eng. A literature review of new types of sequencing batch biofilm reactor). *Zeszyty Naukowe Politechniki Rzeszowskiej, Budownictwo i Inżynieria Środowiska* 268, 56: 67–85.

Masłoń, A. & Tomaszek, J.A. 2009c. Oczyszczanie ścieków w sekwencyjnym reaktorze porcjowym ze złożem ruchomym z porowatym nośnikiem biomasy (Eng. Sewage treatment in a sequencing batch reactor with porous biomass carrier mobile bed). *Gaz, Woda i Technika Sanitarna* 11: 31–35.

Masłoń, A. & Tomaszek, J.A. 2010. Keramzyt w systemach oczyszczania ścieków (Eng. Use of the keramsite in wastewater treatment). *Zeszyty Naukowe Politechniki Rzeszowskiej, Budownictwo i Inżynieria Środowiska* 271, 57(3/10): 85–98.

Masłoń, A., Tomaszek, J.A. 2013. Możliwości zastosowania pylistego keramzytu w aspekcie wspomagania technologii osadu czynnego (Eng. Applications of use of powdered keramsite in aspect improvement of activated sludge technology). *7th National Conference on Science and Training „Progress in Environmental Engineering". Polańczyk, Poland, 19–21 September 2013. Conference proceedings: 16–17.*

May-Esquivel, F., Rios-González, L.J., Garza-García Y., Rodríguez Martínez, J. 2008. Performance of a packed reactor of *Opuntia imbricate* for municipal wastewater treatment. Environmental Research Journal: 2(5), 238–245.

Nicolella. C., van Loosdrecht, M.C.M., Heijnen, J.J. 2000. Wastewater treatment with particulate biofilm reactors. *Journal of Biotechnology* 80: 1–33.

Ødegaard, H., Rusten, B., Westrum, T. 1994. A new moving bed biofilm reactor – Applications and results. *Water Science and Technology* 29(10–11): 157–165.

Ødegaard, H., Gisvold, B., Strickland, J. 2000. The influence of carrier size and shape in the moving bed biofilm reactor. *Water Science and Technology* 41(4–5): 383–391.

Pastorelli, G., Andreottola, G., Canziani, R., de Fraja Frangipane, E., de Pascalis, F., Gurrieri, G., Rozzi, A. 1997. Pilot-plant experiments with moving-bed biofilm reactors. *Water Science and Technology* 36(6): 43–50.

Pastorelli, G., Canziani, R., Pedrazzi, L., Rozzi, A. 1999. Phosphorus and nitrogen removal in moving-bed sequencing batch biofilm reactors. *Water Science and Technology* 40(4–5): 169–176.

Podedworna, J., Żubrowska-Sudoł, M. 2001. Wstępne doświadczenia w usuwaniu azotu i fosforu w sekwencyjnym reaktorze porcjowym ze złożem zawieszonym. *Gaz, Woda i Technika Sanitarna* 11: 398–405.

Podedworna, J., Żubrowska-Sudoł, M. 2006. Możliwość ograniczenia pojemności reaktorów biologicznych poprzez zastosowanie złoża ruchomego (Eng. Possibility of limiting biological reactors' capacity by applying a moving bed). *Gaz, Woda i Technika Sanitarna* 3: 19–22.

Podedworna, J., Żubrowska-Sudoł, M. 2008. Efektywność oczyszczania ścieków komunalnych w reaktorze SBR ze złożem ruchomym. (Eng. The efficiency of wastewater treatment in SBR reactor with moving bed). *Gaz, Woda i Technika Sanitarna* 9: 18–21.

Podedworna, J., Żubrowska-Sudoł, M., Grabińska-Łoniewska, A. 2009. Assessment of the propensity of biofilm growth on newfloat carrier media through process and biological experiments. *Water, Science and Technology* 60(11): 2781–2789.

Rusten, B., Eikebrokk, B., Ulgenes, Y., Lygren E. 2006. Design and operations of the Kaldnes moving bed biofilm reactors. *Aquacultural Engineering* 34: 322–331.

Shin, H-S., Park, H-S. 1991. Enhanced nutrient removal in porous biomass carrier sequencing batch reactor (PBCSBR). *Water, Science and Technology* 23(4–6): 719–728.

Spychała, M., Błażejewski, R., Nawrot, T. 2013. Performance of innoative textile biofilters for domestic wastewater treatment. *Environmental Technology:* 34(1–4), 157–163.

Szilágyi, N., Kovács, R., Kenyeres, I., Csikor, Zs. 2011. Performance of a newly developed biofilm-based wastewater treatment technology. *IWA 1st Central Asian Regional Young Water Professionals Conference. 22–24 September 2011, Almaty, Kazakhstan.*

Szilágyi, N., Kovács, R., Kenyeres, I., Csikor, Zs. 2013. Biofilm development in fixed bed biofilm reactors: experiments and simple models for engineering design purposes. Water, Science and Technology: 68(6), 1391–1399.

Tomaszek, J.A., Masłoń, A. The keramsite as a new support material for the biofilm in the MBSBBR technology. *IWA Specialist Conference "Water and Wastewater Treatment Plants in Towns and Communities of the XXI Century Technologies, Design and Operation". Moscow, Russia, 2–4 June 2010.*

Wiszniowski, J., Surmacz-Górska, J., Robert, D., Weber, J.-V. 2007. The effect of landfill leachate composition on organics and nitrogen removal in an activated sludge system with bentonite additive. *Journal of environmental management:* 85, 59–68.

Valvidia, A., González-Martinez, S., Wilderer, P.A. 2007. Biological nitrogen removal with three different SBBR. *Water, Science and Technology:* 55(7), 245–254.

Yang, S., Yang, F., Fu, Z., Wang, T., Lei, R. 2010. Simultaneous nitrogen and phosphorus removal by novel sequencing batch moving bed membrane bioreactor for wastewater treatment. *Journal of Hazardous Materials*: 175, 551–557.

Zieliński, M., Krzemieniewski, M. 2007. The Effect of Microwave Electromagnetic Radiation on Organic Compounds Removal Efficiency in a Reactor with a Biofilm. *Environmental Technology:* 28(1), 41–47.

Żubrowska-Sudoł, M. 2004. Zastosowanie złoża ruchomego (moving bed) w technologii oczyszczania ścieków (Eng. Use of movable deposit in sewage treatment technology). *Gaz, Woda i Technika Sanitarna* 7–8: 266–269.

Progress in Environmental Engineering – Tomaszek & Koszelnik (eds)
© *2015 Taylor & Francis Group, London, ISBN: 978-1-138-02799-2*

A review of current knowledge on N_2O emissions from WWTPs

J.A. Tomaszek & J. Czarnota
*Department of Environmental and Chemistry Engineering, Rzeszów University of Technology,
Rzeszów, Poland*

ABSTRACT: Analysis of the literature data allows for the identification of the most important factors or operating parameters leading to N_2O emission. Our literature review characterises emissions of nitrous oxide from anthropogenic contaminations, especially from WWTPs: identifies the biological processes (nitrification and denitrification) and influencing factors or operating parameters responsible for N_2O creation (dissolved oxygen, nitrite concentration, COD:N ratio, pH) and seeks to encourage a better understanding of the mechanisms of N_2O formation. This review provides advice and recommends that a strategy to avoid N_2O emission must consider nitrifier denitrification by ammonium-oxidising bacteria, hydroxylamine oxidation as well as heterotrophic denitrification. We also review the N_2O measurement techniques currently in use.

1 INTRODUCTION

Gases that trap heat in the atmosphere are called Greenhouse Gases (GHGs). Carbon dioxide (CO_2) enters the atmosphere through the burning of fossil fuels, solid waste, trees and wood products, and also as a result of certain chemical reactions. Methane (CH_4) is emitted during the production and transport of coal, natural gas and oil, while methane emissions are also generated by livestock, other agricultural practices and the decay of organic waste in municipal landfills for solid waste.

Nitrous oxide (N_2O) is an important greenhouse gas and a major sink for the stratospheric ozone layer when it reacts with atomic oxygen to form nitric oxide (Tallec et al. 2008, Ravishankara et al. 2009, Montzka et al. 2011, Liu et al. 2014). N_2O is emitted into the atmosphere from natural and anthropogenic sources, including the oceans, soil, agricultural activities (fertiliser use), biomass burning and various industrial processes, as well as during the combustion of fossil fuels and solid waste. Anthropogenic sources may account for about 40% of total N_2O emissions (IPCC 2007, Desloover 2012). Even a low emission of N_2O contributes significantly to a WWTP's greenhouse gas footprint (Miąsik et al. 2013). Fluorinated gases: hydrofluorocarbons, perfluorocarbons and sulfur hexafluoride are powerful synthetic greenhouse gases emitted from a variety of industrial processes. Each gas's effect on climate change depends on amounts in the atmosphere (concentration, or abundance). GHG concentrations are measured in ppm, ppb, and even parts per trillion. Each of these gases can remain in the atmosphere for different periods of time, ranging from a few years to thousands of years.

The global warming effect is expressed as a Global Warming Potential (GWP). GWP has been calculated to reflect how long a gas remains in the atmosphere on average, as well as how strongly it absorbs energy. Gases with a higher GWP absorb more energy than those with a lower GWP, and thus contribute more to warming the Earth.

Nitrous oxide is an atmospheric trace gas that has attracted considerable scientific attention. It is a potent greenhouse gas with respective GWP 298 times that of CO_2 over a 100-year time horizon (Cakir 2005, IPCC 2007, Prendez & Lara-Gonzalez 2008, Snip 2010), and it has life time of 114 years (Hu et al. 2010). The N_2O atmospheric concentration in 2008 was estimated at approximately 320 ppb (Hu et al. 2010), which is about 20% higher than during the preindustrial era (Battle et al. 1996, Fluckinger 1999). N_2O is estimated to contribute about 6% of the global radiative forcing due to greenhouse gases (Wuebbles 2009). The mean growth rate of N_2O has been 0.78 ppb per year over the past 10 years (WMO 2009a).

The objectives of this literature review are to characterise emissions of nitrous oxide from WWTPs, to identify the biological processes and influencing factors (operating parameters) that are responsible for its creation, and to encourage a better understanding of the mechanisms behind N_2O formation. We also review N_2O measurement techniques currently in use.

2 NITROUS OXIDE EMISSION FROM ANTHROPOGENIC CONTAMINATION

It is estimated that about two-thirds of overall N_2O are emitted by microbial processes occurring mainly in agriculture, but also in biological wastewater treatment (USEPA 2009).

2.1 Nitrous oxide emissions from agriculture

Emissions of nitrous oxide from agricultural soils contribute significantly to the anthropogenic greenhouse effect (Nyćkowiak et al. 2012). Atmospheric N_2O levels are increasing at the rate of 0.2–0.3% per year, suggesting an anthropogenic N_2O source of about half the size of the natural microbial source from soils and oceans. Enhanced microbial N_2O production associated with agricultural perturbations to the nitrogen cycle in fact accounts for an estimated 75% of this anthropogenic source. Biomass burning, which is closely tied to agriculture, supplies another $\sim$7%, and direct industrial generation of N_2O contributes the remainder (IPCC 1997). Natural and agricultural N_2O sources are not always easy to separate, since both are microbial in nature, and since agricultural N_2O production has to some extent supplanted its natural counterpart (Mosier et al. 1998). The 1997 estimate based on the IPCC Phase II methodology suggests that, out of a total agricultural source of $6.3 \, \text{Tg N} \cdot \text{yr}^{-1}$, $0.9 \, \text{Tg N} \cdot \text{yr}^{-1}$ overlaps with or replaces natural N_2O production in soils and oceans (IPCC 1997).

To assess the global emissions of N_2O from agricultural soils, the total flux should represent N_2O from all possible sources: native soil N, N from recent atmospheric deposition, use of fertiliser in past years, N from crop residues, N from subsurface aquifers below study areas, and current use of N fertilisers. Of these N sources, only synthetic fertilizers and animal manures, as well as the area of fields cropped with legumes are associated with sufficient global data to allow their input as regards N_2O production to be estimated (Monsier et al. 1996). The amount of N_2O derived from the nitrogen applied to agricultural soils via atmospheric deposition or biologically-fixed N, is not known accurately. However, it is estimated that the direct worldwide N_2O emission from agricultural fields as a result of the deposition of all the aforementioned nitrogen sources is 2–3 Tg annually (Monsier 1994). Human-induced changes in N cycling in the soil system have influenced the increases in atmospheric N_2O occurring over the past century, and will help dictate future changes in atmospheric levels of the gas. An unknown, but probably considerable, amount of N_2O is generated by farming activities associated with food production and consumption (Monsier 1994). A growing amount of N_2O emission data are now available for different soil-land-use-systems in various climates. The integration of these data in global and national N_2O budgets is leading to improved estimates. Nevertheless, it is surprising how rare N_2O emission calculations on the meso and macro scales are. Bareth et al. (2001) introduced a new method to help with this, estimating annual N_2O emission potential from agricultural soils in the $775 \, \text{km}^2$ dairy farming region of southern Germany they examined at c. $3.0 \, \text{kg N}_2\text{O-N} \cdot \text{ha}^{-1}$.

Agricultural perturbations to the global nitrogen cycle lead directly or indirectly to enhanced biogenic production of nitrous oxide (N_2O). Direct pathways include microbial nitrification and denitrification of fertiliser- and manure-nitrogen that remains in agricultural soils or animal waste-management systems. In turn, the ondirect emissions category in theory encompasses 5 different sources:

(i) volatilisation and subsequent atmospheric deposition of NH_3 and NO_x,
(ii) nitrogen leaching and runoff,

Table 1. Global N_2O emissions calculated with the IPCC Guadlines [Tg N/yr] (IPCC 1997).

Direct soil emissions	
• subtotal	2.1 (0.4–3.8)

Animal production	
• subtotal	2.1 (0.6–3.1)

Indirect emissions:	
atmospheric deposition	0.3 (0.06–0.6)
nitrogen leaching and runoff	1.6 (0.13–7.7)
human sewage	0.2 (0.04–2.6)
• subtotal	2.1 (0.23–11.9)

Total	6.3 (1.2–17.9)

*values in parentheses indicate estimate range which is derived from the emission factor ranges.

(iii) human consumption of crops followed by municipal sewage treatment,
(iv) formation of N_2O in the atmosphere – from NH_3,
(v) food processing.

In practice, sources (iv) and (v) are not included in the methodology due to a lack of information thereon, this leaving indirect pathways involving nitrogen that is removed from agricultural soils and animal waste-management systems via volatilisation, leaching, runoff, or the harvesting of crop biomass (Table 1) (IPCC 1997).

Several N_2O-flux-measurement techniques have been used in recent agricultural field studies which utilise different chambers along with new analytical techniques in measurement. These studies reveal that it is not the measurement technique that is responsible for much of the uncertainty surrounding the N_2O flux values found in the literature, but rather a real effect reflecting the diverse possible combinations of physical and biological factors capable of controlling gas fluxes (Monsier et al. 1996). There are a variety management techniques which should reduce the amount of N application needed to grow crops and to limit N_2O emissions. Nitrification inhibitors represent an option providing for decreased N fertiliser use, and additionally mitigating N_2O emissions from agricultural soils directly. Inhibitors may be selected for climatic conditions and type of cropping system, as well as the type of nitrogen (solid mineral N, mineral N in solution, or organic waste materials), and then be applied along with fertilisers.

2.2 Nitrous oxide emissions from WWTPs

N_2O can be produced and emitted directly from wastewater treatment systems (Ahn et al. 2010). The Environmental Protection Agency of the United States has shown that N_2O emissions from the WWTP section account for 3% of the emissions from all sources (US-EPA 2006). The Intergovernmental Panel on Climate Change has in turn estimated that N_2O emission from WWTPs account for about 2.8% of total emissions from anthropogenic sources (IPCC 2007). The global N_2O emission from human sewage treatment was estimated at $0.22 \, Tg \cdot yr^{-1}$ for 1990 (Mosier et al. 1999), which is in turn 3.2% of the total estimated anthropogenic N_2O emission. Scientists foresee (anticipate) an increase in nitrous oxide emission averaging about 13% between 2005 and 2020 (Law et al. 2012a).

2.2.1 Mechanisms of N_2O production in the course of biological wastewater treatment

Nitrous oxide is emitted during biological nitrogen removal from wastewater, through autotrophic nitrification and subsequent heterotrophic denitrification.

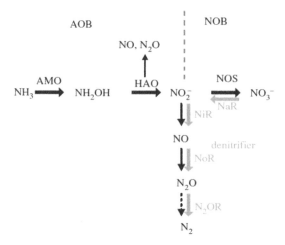

Figure 1. Biological nitrogen conversions (modified from Law et al. 2012a).

Nitrification entails the aerobic oxidation of ammonium (NH_4^+) to nitrate (NO_3^-) via nitrite (NO_2^-), as carried out via a two-step reaction by ammonium-oxidising bacteria (AOBs) e.g. *Nitrosomonas, Nitrosospira* or *Nitrosocystis* which convert NH_4^+ to NO_2^-, as well as by nitrite-oxidising bacteria (NOBs) e.g. *Nitrobacter* and *Nitrospira*, that oxidise NO_2^- to NO_3^-. AOBs and NOBs use ammonia or nitrite as their energy source and CO_2 as the source of carbon (Konneke et al. 2005, Szewczyk 2005). Even though N_2O is not present as an intermediate in the main catabolic pathway of nitrification, AOBs are known to produce it. The process whereby this happens has predominantly been associated with nitrifier denitrification, i.e. the reduction of NO_2^- by AOB in combination with ammonia, hydrogen or pyruvate as electron donors (Kampschreur et al. 2009, Wunderlin et al. 2012).

Although the nitrification step involves both AOBs and NOBs, it is accepted that the latter do not contribute to N_2O production (Law et al. 2012a).

According to Kampschreur et al. (2009), nitrification is performed by three different groups of microorganisms: AOBs, ammonium-oxidising *Archaea* (AOAs) that convert ammonium into nitrite, and NOBs. (In this regard it should be noted that the *Archaea* are a group of microbes resembling bacteria but actually different from them). AOAs are found to occur at WWTPs operating at low dissolved oxygen (DO) levels and with long solid retention times (Park et al. 2006). Ammonium oxidation can also be performed by heterotrophic bacteria (HAOBs). Heterotrophic ammonia oxidation may only prevail over that involving AOBs where the organic load is relatively high (COD:N > 10), while DO is low. Although neither AOAs nor HAOBs play any more significant role in conventional N removal, they might be significant in the production of N_2O (Kampschreur et al. 2009).

Another route also linked to the production of nitrous oxide by AOBs during nitrification is that entailing hydroxylamine (NH_2OH) oxidation. This process follows two stages (Fig. 1).

The first stage called nitritation has two steps. Ammonium is first oxidised to hydroxylamine, the reaction being catalysed by ammonia mono-oxygenase (AMO), which is located in the cell membrane (1) (Szewczyk 2005, Law et al. 2012a).

$$NH_4^+ + O_2 + 2[H] \xrightarrow{AMO} NH_2OH + H^+ + H_2O \qquad (1)$$

Then, during the step two, hydroxylamine is oxidised to NO_2^-. This reaction is catalysed by hydroxylamine oxidoreductase (HAO), which is located in the periplasm (2) (Szewczyk 2005, Law et al. 2012a).

$$NH_2OH + H_2O \xrightarrow{HAO} NO_2^- + H^+ + 4[H] \qquad (2)$$

The second stage of nitrification (nitratation) it is a further oxidation of NO_2^- to NO_3^-. This one-step reaction is catalysed by nitrite oxidoreductase (NOS) (Sadecka 2010, Law et al. 2012a). The production of N_2O via the hydroxylamine route is probably related to highly imbalanced metabolic activity of AOBs (Yu et al. 2010), or else the chemical decomposition of hydroxylamine as well as chemical oxidation with NO_2^- in the role of electron acceptor (chemodenitrification) (Wunderlin 2012). Nitrifying bacteria can only grow if in the constant presence of some dissolved concentration ($\geq 1.5\,mg \cdot L^{-1}$). They are very sensitive to a lower DO concentration. An insufficient supply of DO in a nitrifying process, especially when the concentration of NO_2^- is high, leads to incomplete nitrification. AOBs then reduce nitrite to NO and N_2O (Szewczyk 2005, Hu et al. 2010). A probable mechanism for nitrous oxide formation during hydroxylamine oxidation (as a result of the high activity of HAO) proceeds as follows (Sadecka 2010, Desloover et al. 2012, Law et al. 2012a): $NH_2OH \rightarrow NOH \rightarrow N_2O$.

The concurrent reaction involves: (i) conversion of NH_2OH to a nitrosyl radical (NOH); and then (ii) conversion of NOH to NO_2^-. N_2O and NO can be formed from the activity of HAO through the unstable NOH intermediate. NO is generated as an intermediate during the enzymatic splitting of NOH to NO_2^-, whereas N_2O is produced through the unstable breakdown of NOH. Despite this pathway having been postulated for a long time, its relevance to wastewater treatment processes has not been fully confirmed (Law et al. 2012a). Another route to the formation of N_2O entails biological reduction of nitric oxide, which is generated in the course of hydroxylamine oxidation (Law et al. 2012a). In conditions of anoxia, and in the presence of hydroxylamine or ammonium, *Nitrosomonas* can reduce NO_2^- and produce N_2O (Jetten 1998, Podedworna & Sudoł 2004).

Biological NH_2OH oxidation is hypothesised to contribute to N_2O production mainly at high NH_4^+ and low NO_2^- concentrations in combination with a high nitrogen oxidation rate. The production of N_2O by heterotrophic denitrification is likely to be of minor importance when operated without significant NO_2^- accumulation ($<2\,mg\,N \cdot L^{-1}$) (von Schulthess et al. 1994) and under anoxic conditions. Therefore, to avoid N_2O emissions, biological wastewater treatment plants should be operated at low NH_4^+ and NO_2^- concentrations, which means a high solid retention time, equalisation of load variation (e.g. with digester liquid) and optimal control of sludge recycling depending on COD and NO_3^- loads in the anoxic zone.

Dissimilatory reduction of nitrate (NO_3^-) is the anoxic reduction to dinitrogen gas (N_2) by heterotrophic denitrifiers, coupled with electron transport phosphorylation (Schulthess et al. 1994). Denitrification, a multistage process, is performed by a metabolically very diverse group of microorganisms, bacteria from genera like *Bacillus, Pseudomonas stutzeri, Ps aeroginosa, Ps celis, Achromobacter denitryficans*, as well as *Archaea* which couple organic and inorganic substrates to the reduction of nitrate, nitrite, nitric oxide and nitrous oxide (Mazurkiewicz 2012, Kampschreur et al. 2009). N_2O is an obligate intermediate in the process of heterotrophic denitrification. The first stage (3) relies on the reduction of NO_3^- as catalysed by nitrate reductase (NAR), together with a reduced form of Q-coenzyme (UQH_2). At the second stage (4), nitrite catalysed by nitrite reductase (NIR) is reduced to nitric oxide. At the subsequent stage of reduction (5), NO catalysed by nitric oxide reductase (NOR) produces N_2O, which is next reduced to nitrogen – the final product of denitrification according to reaction (6), catalysed by nitrous oxide reductase (N_2OR), (Szewczyk 2005, Kampschreur et al. 2009). The process takes place in anoxic conditions, in which nitrate serves as the external electron acceptor (nitrate respiration).

$$NO_3^- + UQH_2 \xrightarrow{NAR} NO_2^- + H_2O + UQ \quad (3) \qquad NO_2^- + 2H^+ \xrightarrow{NIR} NO + H_2O \quad (4)$$

$$2NO + 2H^+ \xrightarrow{NOR} N_2O + H_2O \quad (5) \qquad N_2O + 2H^+ \xrightarrow{N_2OR} N_2 + H_2O \quad (6)$$

Under the typical denitrifying conditions found in the circumstances of the biological wastewater treatment process, NO and N_2O reductases have higher maximum nitrogen turnover than NO_3^- and NO_2^- reductases (von Schulthess 1994). This indicates that N_2O could be completely reduced under anoxic/anaerobic conditions, without any accumulation or emission of the gas taking place.

However, fluctuations in environmental conditions have been found to inhibit N_2OR, with the result that an accumulation of N_2O nevertheless takes place (Law et al. 2012a). The result is incomplete denitrification, with N_2O being generated as the end product instead of N_2. The presence of oxygen can inhibit denitrification enzymes, particularly N_2OR, this leaving DO as key in determining the metabolic mechanisms that trigger N_2O production from either nitrifying or denitrifying microorganisms, depending on whether conditions are aerobic and/or anoxic (Rassamee et al. 2011). This is more relevant in situations where transient DO levels are frequent (Kampschreur et al. 2008a), thereby creating a favourable environment for anoxic and aerobic conditions to co-exist (Aboobakar et al. 2013). Many denitrifying microorganisms are facultative denitrifiers, which preferentially use oxygen as an electron acceptor due to the higher energy field. Some microorganisms can denitrify under both aerobic and anoxic conditions, a process known as aerobic denitrification. Often these microorganisms can also catalyse heterotrophic nitrification (Kampschreur et al. 2009). AOBs can also denitrify from nitrite to N_2O, with ammonium or hydrogen or pyruvate as electron donors. This process known as nitrifier denitrification (described in the nitrification paragraph) takes place under oxygen-limiting conditions or in circumstances of elevated nitrite concentration, and represents one of the main routes leading to N_2O production.

Autotrophic denitrification at the first step oxidises ammonium to nitrite, and next via nitrogen oxides to nitrogen gas (7) (Kampschreur et al. 2009, Miksch & Sikora 2010).

$$NH_3 \rightarrow NO_2^- \rightarrow NO \rightarrow N_2O \rightarrow N_2 \tag{7}$$

The process is led by autotrophic ammonia-oxidising bacteria, e.g. *Nitrosomonas europaea*, at low dissolved oxygen concentrations (Podedworna & Żubrowska-Sudoł 2004, Miksch & Sikora 2010). Hitherto-existing investigation of this process shows that, in circumstances of limited DO concentration in treated wastewater and an exchange from aerobic to anaerobic conditions, the production of N_2O may be considerable, accounting for up to 3.9% of the total nitrification (Podedworna & Żubrowska-Sudoł 2004).

The literature on denitrification also gives information about a possibility of N_2O being produced from nitrite by heterotrophic microorganisms like *Thiosphaera pantotropha*, *Alcaligenes faecalis* and *Pseudomonas aeruginosa*, with an assumption of denitrifying enzyme activity in aerobic conditions. The process is called heterotrophic aerobic denitrification (van Niel et al. 1992, Szewczyk 2005). The generation of N_2 via this process gained confirmation in work by Carter et al. (1995), and Robertson et al. (1995). However, Otte et al. (1996) showed by their experiments that *Alcaligenes faecalis* can produce almost a mole-equivalent quantity of N_2 and N_2O. It was also shown that alternately aerobic and anaerobic conditions cause a 25% decrease of nitrite conversion into N_2O.

2.2.2 *Factors influencing N_2O emission*

Factors leading to N_2O emission from WWTPs have thus far remained within the spheres of discussion and detailed investigations. Some studies conclude that abiotic or abiological N_2O formation play a negligible role in WWTPs (Kampschreur et al. 2008a). However, most work proves that N_2O production in the course of biological wastewater treatment is significant (Aboobakar et al. 2013, Daelman et al. 2013, Wang et al. 2012, Quan et al. 2012).

Among operating parameters playing an important role in influencing N_2O emission is dissolved oxygen (DO). Improved nutrient removal from wastewater as a result of stringent standards has increased the emission of N_2O into the atmosphere, due to the introduction of anoxic zones and low dissolved oxygen concentrations at WWTPs. The emission of NO is reported less frequently, but is expected to follow the same trend as for N_2O, as it is produced via the same biological pathways. Overall, a thorough investigation into the potential contribution of WWTPs to NO and N_2O emissions, and the role of the individual processes in the emissions seems to be lacking.

Emission of nitrous oxide is linked closely with the aeration demand required to keep adequate DO conditions in a nitrifying activated sludge process. Lowering the high costs of aeration in nitrification reactors (about 55% of the total energy consumption of a WWTP) (Aboobakar et al.

2012) through supplies of DO that are insufficient can lead to incomplete nitrification and thus to N_2O emission. A high output of N_2O was observed where DO concentrations were below $2\,mgO_2 \cdot L^{-1}$, especially across the range 0.5 to $1\,mgO_2 \cdot L^{-1}$ (Kampschreur et al. 2008b). At concentrations of DO below $1\,mgO_2 \cdot L^{-1}$, production of N_2O can amount to about 10% of the nitrogen load (Goreau et al. 1980). The highest values for N_2O were observed when oxygen concentrations ranged between 0.1 and $0.3\,mgO_2 \cdot L^{-1}$ (Chuang et al. 2007, Law et al. 2012a). Some investigations to increased N_2O emission resulting in increased aeration during the activated sludge process (Sümer et al. 1995). DO concentration is an important factor influencing N_2O emissions, not only with nitrification but also with denitrification (Otte et al. 1996). The presence of DO during denitrification inhibits the N_2OR enzyme resulting in N_2O emission. Transient differences in DO concentrations in both the aerobic and anaerobic zones can be conductive to N_2O emission (Zheng et al. 1994, Kampschreur et al. 2009). A lower concentration of dissolved oxygen (anoxia conditions) in the nitrification chambers was shown to result in increased loading of organic contamination or restricted oxygen capacity, this in turn giving increased emissions of N_2O (Butler et al. 2009, Foley et al. 2009).

Excessively intensive aeration in nitrification chambers can cause penetration of DO into denitrification chambers and thus an increased emission of N_2O from these parts of a WWTP (Otte et al. 1996, Law et al. 2012a).

A further important factor influencing N_2O emission is nitrite. Being the obligate intermediate of autotrophic nitrification and heterotrophic denitrification, nitrite influences N_2O emissions significantly (Ali et al. 2013). An increase in the nitric concentration during nitrification may be caused by: insufficient aeration in the nitrifying chamber, the presence of toxic compounds, low temperature or a high concentration of ammonium. In consequence, concentrations of nitrite can increase in the denitrifying zone, causing a lowered efficiency of denitrification and the accumulation of NO and N_2O (Colliver & Stephenson 2000, Kampschreur et al. 2009). Tallec et al. (2006) in their research on the introduction of 'artificial' nitrite into the system demonstrated that the emission of N_2O depends to a great extent on nitrite concentration and increases even 8-fold. An influence of nitrite on N_2O emission was also shown in the papers from Sümer et al. (1995) and Alinsafi et al. (2008), Osada et al. (1995) and Ali et al. (2013).

The COD:N ratio is a further factor influencing N_2O emission at WWTPs. The availability of biodegradable organic compounds ($C_{b.org}$) is an essential element which controls N_2O production during the denitrification process. Frequently a limited availability of $C_{b.org}$ causes accumulation of intermediate products of reduction such as NO_2^- or N_2O, and as a consequence N_2O emission (Chung & Chung 2000, Kampschreur et al. 2008b, Law et al. 2012a). Full denitrification takes place when the COD:N ratio is above 4, while for lower values an incomplete denitrification occurs, characterised by during denitrification enzymes (NAR, UQH_2, NIR, NOR and N_2OR) competing for electrons (Law et al. 2012a). The investigations included a determination of the influence of different COD:N ratios (from 1.5 to 4.5) on N_2O emissions, it being observed that – for the lowest value of the COD/N ratio equal to 1.5 – up to 10% of the nitrogen load is emitted as N_2O. This compares with the situation where the value of the ratio is 3.5 and the percentage is then in the 20–30% range (Hanaki et al. 1992, Itokawa et al. 2001). Addition of an external $C_{b.org}$ source is conducive to the elimination of electron competition and hence to an immediate decrease in N_2O emissions. Park et al. (2000) used methanol as an external $C_{b.org}$ source, and obtained a decrease in N_2O emissions from 4.5% to 0.2% of the amount of nitrogen. Alinsafi et al. (2008) carried out investigations on the laboratory scale and noted the highest emission of N_2O during the denitrification process when $COD:NO_3$-N was equal to 3 and the highest concentration of nitrite was equal to $20\,mg\,N \cdot L^{-1}$. These results contradict those of Ahn et al. (2010), who noted low N_2O emission and no relationship between the magnitude of N_2O emissions and the availability or shortage of organic carbon, in both anoxic and aerobic conditions tested at full scale.

Another parameter influencing N_2O emission is pH. In WWTPs the optimum pH range is from 6.5 to 8.0 (Hartmann 1996), while the interval known to be optimal for the denitrification process is in the range 7–8 (Szewczyk 2005). Production of nitrous oxide during nitrification is dependent on pH. In the case of *Nitrosomonas europaea* bacteria, the maximum N_2O emission was determined at a pH of 8.5, as compared with a minimum at pH 6 (Hynes & Knowles 1984). Law et al. (2012b) were also able to note increased N_2O production at a pH of 8, in the case of AOB microorganisms. During the denitrification process a pH value <7 was seen to cause enlargement of the fractions of N_2O and NO (Szewczyk 2005). Conversely, in the case of the denitrification process, Thoern and Soerensson (1996) were able to observe N_2O formation below a pH of 6.8. Similarly, Hanaki et al. (1992) reported an increase in the N_2O emission in a denitrification chamber as the pH was lowered from 8.5 to 6.5. However, as the pH at WWTPs is actually rather stable (usually in the range 7–8), the effect of this factor on N_2O emission is not expected to be major.

Some factors might not be directly related to N_2O emissions but cause changes of parameters that lead to N_2O production. For example short SRT, low temperatures, high salinity and increased ammonium concentration may have resulted in nitrite concentration that induces N_2O emission (Kampschreur et al. 2009).

Most studies of N_2O emissions focus solely on the biological part of WWTPs, with an activated sludge process whereby nitrification and denitrification takes place. However, there are other places at WWTPs in which N_2O emission can occur. Other possible installations are: grit tanks, pre-sedimentation tanks, secondary clarifiers, sludge storage tanks and anaerobic digesters (Prendez & Lara-Gonzalez 2008). However, for correctly operating compartments, and hence in the presence of a nitrogen removal process, the emission of N_2O should not be great (Kampschreur et al. 2009).

2.2.3 *Discussion of results for N_2O emission from different WWTPs*

In estimating N_2O emissions from WWTPs we can proceed along the lines of the International Panel on Climate Change (IPCC), according to whom – from 2006 onwards – the standard N_2O emission factor was down from 1% to 0.5% of nitrogen content in the effluent from a treatment plant (IPCC 2006). The 0.5–1% IPCC factors are based on data for emissions from soils. For countries with advanced centralized WWTPs, a factor for the direct emission from WWTPs is applied, which is $3.2 \, gN \cdot person^{-1} \cdot year^{-1}$, corresponding to approximately 0.035% of N_2O emission of the nitrogen load of a WWTP (Kampschreur et al. 2008b, Kampschreur et al. 2009, Miąsik et al. 2013). In the literature and in IPCC reports there are major discrepancies in valuations of the levels of N_2O emission from wastewater treatment objects, varying across the range 0.001–90% and 0.05–25% of the nitrogen load respectively (ICPP 2006, Kampschreur et al. 2009). The German estimation of the factor for the direct emission of N_2O from wastewater treatment, excluding any influence due to industrial effluent, averages $7.0 \, g \, N_2O \cdot person^{-1} \cdot year^{-1}$ (Thomsen & Lyck 2008). On the basis of an analysis performed in The Netherlands, the value of this same factor, again excluding the influence of industrial wastewaters is equal to $3,2 \, g \, N_2O \cdot person^{-1} \cdot year^{-1}$ (Czepiel et al. 1995). In contrast, in the USA, the index value taking account of the N-load due to industrial effluents averages at $4.0 \, g \, N_2O \cdot person^{-1} \cdot year^{-1}$ (Thomsen & Lyck 2008, Miąsik et al. 2013). According to Mosier et al. (1999), the 1990 level of global emissions of nitrous oxide from domestic wastewater was at $0.22 \cdot 10^6 \, Mg \, N \cdot year^{-1}$, this constituting 3.2% of the total anthropogenic emission of N_2O.

On the basis of the review of the literature conducted we can state that the amount of N_2O-N emitted from WWTPs in relation to the overall amount of nitrogen in treatment plant influent oscillates within the range 0–25%. Such major disparities in emission values can result from different technology and technological conditions, as well as methods of monitoring and quantitative analysis (Law et al. 2012a).

Below, we introduce several examples of different wastewater treatment technologies, together with the results of N_2O emission measurements. Moreover, quantities of N_2O emissions from WWTPs working both on the laboratory and full scales are as presented in Table 2.

The emission of N_2O from a WWTP working as an SBR changed from 2.36 to $8.64 \, g/T$ of wastewater/year, while more than 99% of the emission occurred during the filling and aeration phases (Sun et al. 2013). It was also stated that about 5.6% of the nitrogen supplied to wastewater was converted into N_2O and emitted to the atmosphere, giving a total emission equal to $120 \, T \cdot year^{-1}$. The main factor influencing N_2O production was a low DO concentration during the nitrification process. This information was confirmed by de Mello et al. (2013). During their investigations, at the full-scale of the wastewater treatment plant with activated sludge, emission of nitrous oxide

Table 2. N$_2$O emission factors reported for several lab-scale and full-scale WWTP (Modify from Kampschreur et al. 2009).

Reference	Scale	Type of WWTP	N$_2$O emissions (% of N-load)	Remarks
Czepiel et al. (1995)	full-scale	activated sludge plant	0.035–0.05%	• wastewater type: municipal, • weekly grab samples for 15 weeks, • N$_2$O was emitted in aerated areas, low N$_2$O flux at non-aerated areas.
Sümer et al. (1995)	full-scale	activated sludge plant	0.001%	• wastewater type: municipal, • 2-weekly grab over 1 year, • N$_2$O emission increased with NO$_2$ and NO$_3$ concentrations.
Tallec et al. (2006)	lab-scale	nitrifyng activated sludge	0.1–0.4% of oxidised ammonia	• wastewater type: municipal, • N$_2$O emission is largest at 1 mgO$_2$/L and lower above and below this oxygen concentration – emission increases with nitrite concentration.
Kampschreur et al. (2008a)	full-scale	nitration – anammox sludge water treatment	2.3%	• on-line measurements during 4 days, • N$_2$O emissions increased with decreasing oxygen concentration (aerated stage) and increasing nitrite concentration (anoxic stage).
Foley et al. (2009)	full-scale	7 activated sludge plants with different N/DN configurations	0.06–25.3 (of N denitrified)	• wastewater type: municipal, • lower N$_2$O production observed in plants with low total nitrogen effluent concentrations, • high NO$_2$ concentration led to high N$_2$O emissions.
Ahn et al. (2010)	full-scale	12 BNR plants	0.003–2.59%	• wastewater type: municipal, • online measurement, • aerobic zones contributed substantially more to N$_2$O fluxes than anoxic zones.
Ahn et al. (2010)	full-scale	12 BNR plants	0.003–2.59%	• wastewater type: municipal, • online measurement, • aerobic zones contributed substantially more to N$_2$O fluxes than anoxic zones.
Rajagopal & Beline (2011)	full-scale	1-stage nitritation–denitritation	0.07–0.15%	• wastewater type: manure – piggery wastewater – digestate, • grab every 0.5–1 h/2 × 12 h, • aerobic period was the main source of N$_2$O emissions.
Quan et al. (2012)	lab-scale	aerobic granular sludge	2.2–8.2%	• wastewater type: municipial • N$_2$O emission decreased with decrease of COD/N, • N$_2$O emission decreased with increase in DO level.

took place in both the aerobic and anaerobic phases, albeit being at a much higher level during the aeration phase. The calculated diurnal coefficient or factor for N_2O emission (in the aeration phase) was $2.4 \cdot 10^{-2}$ g $N_2O \cdot person^{-1} \cdot d^{-1}$. The emission accounted for 0.10% of supplied total nitrogen load, and the emission coefficient was thus 3 times higher than that suggested by the IPCC, which is equal to $0.9 \cdot 10^{-2}$ g $N_2O \cdot person^{-1} \cdot d^{-1}$ (3.2 g $N_2O \cdot person^{-1} \cdot year^{-1}$). Hu et al. (2010) applied increased values of aeration in the SBR (1.0; 2.7 and 4.4 $L_{air} \cdot L_{reactor}^{-1} \cdot h^{-1}$) and observed a decrease in N_2O emission for both the anoxic phase (0.23; 0.22 and 0.22 mg N_2O) and the aerobic phase (11.51; 5.02 and 2.92 mg N_2O). The total amount of N_2O emitted to the atmosphere decreased, together with an increase in delivered air, and was equal to 26.1, 8.0% and 5.3% respectively.

One of the new technologies as regards biological wastewater treatment is that involving aerobic granular sludge. Spatial structure of the granules can contribute to incomplete denitrification leading to N_2O emission (Quan et al. 2012). There latter authors made it clear that emission of N_2O depends on intensity of aeration. Investigations were carried out on the laboratory scale in 3 identical SBR reactors (GSBR), but with different aeration intensities (of 0.2; 0.6 and 1.0 $L_{air} \cdot min^{-1}$), working in the successive stages with a changing COD:N ratio (1:0.22; 1:0.15 and 1:0.11). Nitrous oxide emissions were respectively: 8.2%, 6.1% and 3.8%, for COD:N ratios of 1:0.22; 7.0%, 5.1% and 3.5% for the COD:N ratios of 1:0.15 and 4.4%, 2.9%, 2.2% for the COD:N ratio of 1:0.11 influenced nitrogen load. In turn, Rathnayake et al. (2013) determined average emission of N_2O from the GSBR at the level 0.32 ± 0.17 mg $N \cdot L^{-1} \cdot h^{-1}$, or $0.8 \pm 0.4\%$ of the supplied nitrogen load.

The production of N_2O in full-scale reject water treatment is largely unknown. However, due to the high volumetric nitrogen conversion rates, significant emissions can be expected. The studies on the impact of dynamic process conditions on the emission of N_2O during reject water treatment by the nitrification-anammox process in the two reactors showed that 2.3% of the nitrogen load was converted into N_2O. The greater part of the N_2O (78%) was produced in the nitritation reactor, by AOBs in the course of the denitrification process (Kampschreur 2008b).

2.2.4 *Analytical techniques for the measurement of N_2O*

While the last few years have brought a distinct intensification of problems connected with the emission of nitrous oxide to the atmosphere, great progress has also been made with measurements of N_2O. However, there remains a necessity for research in this field to be improved. Investigations are in turn tending towards advanced and simultaneously inexpensive methods by which to measure N_2O over a wide area, with account taken of the spatial variation to N_2O emissions across longer time intervals. The World Meteorological Organization's Global Atmosphere Watch (WMO-GAW) recommends an accuracy and precision to measurements of N_2O of less than 0.1 ppb. (WMO 2009b). Measurements of N_2O are difficult because of (i) the low (320 ppb) concentration of the gas in the atmosphere, which is below the detection threshold of many analytical techniques, (ii) irregularity of fluxes, which show major temporal and spatial variations, due to the multiplicity of processes that both produce and consume the gas (Dalal et al. 2003, Rapson & Dacres 2014).

A comprehensive overview of sampling methods and analytical techniques (used already or in the research sphere) is provided in Tables 3 and 4.

Each of these sampling methods and analytical techniques has its advantages and disadvantages. The progress of the last few years affords possibilities for very low concentrations of the gas N_2O to be measured. The emission of nitrous oxide from WWTPs is generally detected through immediate measurement of N_2O emissions in the gaseous phase. The gas samples are collected from closed chambers. The analytical technique used most widely in the measurement of nitrous oxide is gas chromatography (GC), as equipped with an electron-capture detector (ECD), (de Mello et al. 2013, Alinsafi et al. 2008, Mao et al. 2006, Ikeda-Ohtsubo et al. 2013). The major disadvantage of the GC technique is the time taken to run samples. If the GC system is linked to an isotope-ratio mass spectrometer (IRMS), then it is possible for isotopic analysis of N_2O to be carried out (Tomaszek et al. 2008). Competition for the GC technique is provided by the optical techniques. One of the most successful of these is cavity ring-down spectroscopy (CRDS). Using this technique, a gas sample

Table 3. Sampling methods.

Method	Characteristic	
Chamber methods	• Chamber (manual or automated) is placed over the examined surface, closed for a specified period of time during which, samples are collected and next analyzed for a concentration of N_2O. • Size of the chamber from less than $1\,m^3$ to greater than $150\,m^3$. • Examined area for the automated chambers $<25\,m^2$. • Advantages: easy to use and do not require extremely accurate analytical techniques. • Disadvantages:disturbance of the microclimate and therefore a limited time of detection (Turner et al. 2008, Rapson & Dacres 2014).	
Micrometeorological methods	• In this techniques gas sensors are placed on towers that measure wind, temperature and concentration of gas. • Advantages: utilized to measure gas fluxes over a large area ($1–10\,km^2$), (Dalal et al. 2003, Rapson & Dacres 2014).	• *Eddy covariance* – direct measuring of the rate of vertical transport of the gas (the flux over a surface) (Pattey et al. 2007, Denmead 2008). • *Eddy accumulation* – sampling technique, where air is collected into two separate containers at the rate proportional to the vertical wind speed (Denmead 2008, Rapson & Dacres 2014). • *Flux-gradient methods* – the vertical flux is determined by measuring gas concentration at two or more different heights and recording the horizontal wind speed what simplified the measurements (Denmead 2008). • *Integrated horizontal flux (IHF)* – the profiles of the horizontal wind speed and gas concentrations are made at the center of the test size (Denmead et al. 1998). • *Backward Lagrangian stochastic (bLs) dispersion techniques* – uses Lagrangian model of air flow to calculate surface fluxes from the small area, using measurements of wind speed, wind direction and the gas concentration downwind (Rapson & Dacres 2014). • *Moving platforms* – e.g. ships, trains – moving platforms allow measurements in both horizontal and vertical directions with high spatial and temporal resolution (Rapson & Dacres 2014).

is analyzed in a high-finesse optical cavity, the optical absorbance of the sample being determined by reference to the light dissipation rate in the optical cavity, in this way typically providing parts-per-billion mixing ratio or isotopic ratio measurements of a particular gas species of interest, which offer a good approximation independent of intensity fluctuations characterising the excitation light source. The technique has been implemented successfully in greenhouse gas analysers. The CRDS metod of detection is highly sensitive, down to a precision of 0.05 ppb at 1 Hz (Rapson & Dacres 2014). Such measurements of extremely low concentrations of N_2O thus represent a great advance in research seeking to better elucidate the transformations and emissions nitrous oxide is subject to. On account of the high solubility of nitrous oxide in water, frequent use of micro-sensors for N_2O is made to determine amounts of the gas dissolved in wastewater (Kampschreur et al. 2008a, Foley et al. 2009, Zhang et al. 2012). This means of on-line measurement, as compare with gas detection methods, provides the possibility for levels of emission to be determined, and N_2O generation in bioreactors calculated (Quan et al. 2012).

Table 4. Analytical techniques for measuring N_2O (Modify from: Rapson & Dacres 2014).

Method	Sensitiv	Advantages of technique	Disadvantages of technique
I. CHROMATOGRAPHIC TECHNIQUES Gas chromatography (GC) with an electron – capture detector (ECD)	30 ppb LOD Precision: 0.18–0.4 ppb	• lower cost, • widely used, allowing easy data comparison, • if linked to IRMS, isotope analysis can be carried out,	• non-continuous, • frequent calibration required, • drift means reference needs to be run every 3 samples, • long run time (5 min),
II. OPTICAL TECHNIQUES			
1. FTIR (Fourier-transform infrared spectroscopy	Precision: 0.1 ppb (1 min) and 0.03 ppb (10 min)	• continuous measurements, • portable, • lower calibration requirements, • broadband spectrum allowing the measurement of multiple components and spectra can be reanalyzed,	• high cost of instruments, • low brightness of light source, • slower measurements compared to lasers, • analysis of data can be more complicated,
2. Infrared laser absorption spectroscopy:	Precision:	• allows rapid and highly sensitive measurements,	• cryogenic cooling required • narrowband, can only measure a single species or pair per laser,
o Lead-salt diode laser	<1 ppb in 5 sec.	• lower interference from other trace gases,	• low pressure required, • expensive,
o Quantum-casade laser (QCLs)	0.05 ppb in 1 Hz	• no cryogenic cooling required making lasers more portable, • higher power than lead salt lasers giving a higher signal to noise ratio and faster measurements, • can be linked with high finesse optical cavities such as CRDS and ICOS, • carry out isotopic analysis without pre-concentration,	• spectral quality not as high as Lead Salt Lasers, • narrowband, can only measure a single, • species or pair per laser, • low pressure required, • expensive,
III. AMPEROMETRIC METHODS Microsensor	22 ppb LOD	• low cost of instruments, • extremely portable,	• sensor drift, therefore not suitable for long term monitoring, • only dissolved N_2O can be measured,
IV. NEW METHOD UNDER DEVELOPMENT			
1. Amperometric biosensor	0.1 mM (4.4 ppm) – – 8 mM (352 ppm)		• in the very early stages of development and the sensitivity obtained is unsuitable for N_2O-flux measurements at present,
2. Optical fibers	0.2–1.8 ppb	• for in-situ analysis, • response time is faster than that obtained using SPME-GC-MS,	• may be too sensitive for measurement in places where levels N_2O is higher • for measuring N_2O fluxes from soils as a faster alternative to GC methods (low detection range),
3. Modified-SnO_2 surfaces	detection of 10–300 ppm N_2O at 500°C	• sensors could be used to detect N_2O in the workplace, or possibly used with closed chambers,	–

3 CONCLUSIONS

In summarising our literature review we can conclude that the most important factors influencing the N_2O emissions are:

- a low concentration of dissolved oxygen at the nitrification stage,
- a concentration of dissolved oxygen in the denitrification chamber higher than established for the steady-state conditions in the anoxic zone,
- an increased concentration of nitrite at both the nitrification and denitrification stages,
- low COD:N ratio at the denitrification stage.

Moreover, the emission of N_2O during wastewater treatment is related to other operating conditions such as: pH, temperature, genus of microorganisms present and timing of operations.

The emissions of nitrous oxide from WWTPs is relatively limited (accounting for about 3% of total anthropogenic N_2O emission). The relevant data show a large variation in the fraction of nitrogen emitted as N_2O, and it remains unclear whether nitrifying or denitrifying microorganisms are the main source of N_2O emissions.

This literature review provides evidence that strategies to avoid N_2O emissions must consider nitrifier denitrification by AOBs, hydroxylamine oxidation, as well as heterotrophic denitrification.

REFERENCES

Aboobakar, A., Cartmell, E., Stephenson, T., Jones, M., Vale, P., Dotro, G. 2013. Nitrous oxide emissions and dissolved oxygen profiling in a full-scale nitrifying activated sludge treatment plant. *Water Research* 47: 524–534.

Ahn, J.H., Kim, S.P., Park, H.K., Rahm, B., Pagilla, K., Chandran, K. 2010. N_2O Emissions from activated sludge processes, 2008–2009: results of a national monitoring survey in the United States. *Environmental Science and Technology* 44: 4505–4511.

Ali, T.U., Ahmed, Z., Kim, D.J. 2013. Estimation of N_2O emission Turing wastewater nitrification with activated sludge: Effect of ammonium and nitrate concentration by regression analysis. *Journal of Industrial and Engineering Chemistry* 1637: 1–6.

Alinsafi, A., Adouani, N., Beline, F., Lendormi, T., Limousy, L., Sire, O. 2008. Nitrite effect on nitrous oxide emission from denitrifying activated sludge. *Process Biochemistry* 43: 683–689.

Bareth, G., Heincke, M., Glatzel S. 2001. Soil-land-use-system approach to estimate nitrous oxide emissions from agricultural soils. *Nutrient Cycling in Agroecosystems* 60: 219–234.

Battle, M., Bender, M., Sowers, T., Tans, P.P., Butler, J.H., Elkins, J.W., Conway, T., Zhang, N., Lang, P., Clarke, A.D. 1996. Atmospheric gas concentrations over the past century measured in air from firn at South Pole. *Nature* 383: 231–235.

Butler, M.D., Wang, Y.Y., Cartmell, E., Stephenson, T. 2009. Nitrous oxide emissions for early warning of biological nitrification failure in activated sludge. *Water Research* 43: 1265–1272.

Cakir, F.Y., Stenstrom, M.K. 2005. Greenhouse gas production: A comparison between aerobic and anaerobic wastewater treatment technology. *Water Research* 39: 4197–4203.

Carter, J.P., Hsiao, Y.H., Spiro, S., Richardson, D.J. 1995. Soil and sediment bacteria capable of aerobic nitrate respiration. *Applied and Environmental Microbiology* 61: 2852–2858.

Chung, Y.C., Chung, M.S. 2000. BNP test to evaluate the influence of C/N ratio on N_2O production in biological denitrification. *Water Science and Technology* 42: 23–27.

Chuang, H.-P., Ohashi, A., Imachi, H., Tandukar, M., Harada, H. 2007. Effective partial nitrification to nitrite by down-flow hanging sponge reactor under limited oxygen condition. *Water Res.* 41: 295–302.

Colliver, B.B., Stephenson, T. 2000. Production of nitrogen oxide and dinitrogen oxide by autotrophic nitrifiers. *Biotechnology Advances* 18(3): 219–232.

Czepiel, P., Crill, P., Harriss., R. 1995. Nitrous oxide emissions from municipal wastewater treatment. *Environmental Science and Technology* 29(9): 2352–2356.

Daelman, M.R.J., De Beats, B., van Loosdrecht, M.C.M., Volcke, E.I.P. 2013. Influence of sampling strategies on the estimated nitrous oxide emission from wastewater treatment plants. *Water Research* 47: 3120–3130.

Dalal, R.C., Wang, W.J., Robertson, G.P., Parton, W.J. 2003. Nitrous oxide emission from Australian agricultural lands and mitigation options. *Australian Journal of Soil Research* 41: 165–195.

de Mello, W.Z., Brotto, R.P., Kligerman, D.C., Oliveira A. 2013. Nitrous oxide emissions from an intermittent aeration activated sludge system of an urban wastewater treatment plant. *Quim. Nova* 36(1): 16–20.

Denmead, O.T., Harper, L.A., Freney, J.R., Griffith, D.W.T., Leuning, R., Sharpe, R.R. 1998. A mass balance method for non-intrusive measurements of surface-air trace gas exchange. *Atmospheric Environment* 32: 3679–3688.

Denmead, O.T. 2008. Approaches to measuring fluxes of methane and nitrous oxide between landscapes and the atmosphere. *Plant Soil* 309: 5–24.

Desloover, J., Vlaeminck, S., Clauwaert, P., Verstraete, W., Boon, N. 2012. Strategies to mitigate N_2O emissions from biological nitrogen removal systems. *Biotechnology* 23: 474–482.

Fluckiger, J., Dallenbach, A., Blunier, T., Stauffer, B., Stocker, T.F., Raynaud, D., Barnola, J.M. 1999. Variations in atmospheric N_2O concentration during abrupt climatic changes. *Science* 285: 227–230.

Foley, J., de Haas, D., Yuan, Z., Lant, P. 2009. Nitrous oxide generation in full-scale biological nutrient removal wastewater treatment plants. *Water Research* 44: 831–844.

Goreau, T.J., Kaplan, W.A., Wofsy, S.C., McElroy, M.B., Valois, F.W., Watson, S.W. 1980. Production of NO_2 and N_2O by nitrifying bacteria at reduced concentrations of oxygen. *Appl. Environ. Microbiol.* 40: 526–532.

Hanaki, K., Hong, Z., Matsuo, T. 1992. Production of nitrous oxide gas during denitrification of wastewater. *Water Science and Technology* 26(5–6): 1027–1036.

Hartmann, L. 1996. Biologiczne oczyszczanie ścieków. Wyd. Instalator Polski. Warszawa 1996.

Hu, Z., Zhang, J., Li, S., Xie, H., Wang, J., Zhang, T., Li, Y., Zhang, H. 2010. Effect of aeration rate on the emission of N_2O in anoxic–aerobic sequencing batch reactors (A/O SBRs). *Journal of Bioscience and Bioengineering* 109(5): 487–491.

Hynes, R.K., Knowles, R. 1984. Production of nitrous oxide by Nitrosomonas europaea: effects of acetylene, pH, and oxygen. *Canadian Journal of Microbiology* 30(11): 1397–1404.

Ikeda-Ohtsubo, W., Miyahara, M., Kim, S.W., Yamada, T., Matsuoka, M., Watanabe, A., Fushinobu, S., Wakagi, T., Shoun, H., Miyauchi, K., Endo, G. 2013. Bioaugmentation of a wastewater bioreactor system with the nitrous oxide-reducing denitrifier Pseudomonas stutzeri strain TR2. *Journal of Bioscience and Bioengineering* 115(1), 37–42.

IPCC. 1997. Intergovernmental Panel on Climate Change/Organization for Economic Cooperation and Development. *Guidelines for National Greenhouse Gas Inventories*. OECD/OCDE, Paris.

IPCC. 2006. Guidelines for National Greenhouse Gas Inventories. In: Eggleston, H.S., Buendia, L., Miwa, K., Ngara, T., Tanabe, K. (Eds.). IGES, Japan.

IPCC. 2007. Climate change 2007: synthesis report. Contribution of Working Groups I, II and III to the Fourth Assessment Report of the Intergovernmental Panel on Climate Change (eds C. W. Team, R. K. Pachauri, A. Reisinger). Geneva, Switzerland: IPCC.

Itokawa, H., Hanaki, K., Matsuo, T. 2001. Nitrous oxide production in high-loading biological nitrogen removal process under low COD/N ratio condition. *Water Research* 35: 657–664.

Jetten, M.S.M., Hom, S.J., van Loosdrecht, M.C.M. 1997. Towards a more sustainable municipal wastewater treatment system. *Water Science and Technology* 35: 171–180.

Kampschreur, M.J., Tan, N.C.G., Kleerebezem, R., Picioreanu, C., Jetten, M.S.M., van Loosdrecht, M.C.M. 2008a. Effect of dynamic process conditions on nitrogen oxide emission from a nitrifying culture. *Environmental Science & Technology* 42: 429–435.

Kampschreur, M.J., van der Star, W.R.L., Wielders, H.A., Mulder, J.W., Jetten, M.S.M., van Loosdrecht, M.C.M. 2008b. Dynamics of nitric oxide and nitrous oxide emission during full-scale reject water treatment. *Water Research* 42(3): 812–826.

Kampschreur, M.J., Temmink, H., Kleerebezem, R., Jettena, M.S.M., van Loosdrecht M.C.M. 2009. Nitrous oxide emission during wastewater treatment. *Water Research* 43: 4093–4103.

Konneke, M., Bernhard, A.E., de la Torre, J.R., Walker Christopher, B., Waterbury, J.B., Stahl, D.A. 2005. Isolation of an autotrophic ammonia-oxidizing marine archaeon. *Nature* 437(7058): 543–546.

Law, Y., Ye, L., Pan, Y., Yuan, Z. 2012a. Nitrous oxide emissions from wastewater treatment processes. *Philosophical Transactions of the Royal Society B* 367: 1265–1277.

Law, Y., Lant, P., Yuan, Z. 2012b. The effect of pH on N_2O production under aerobic conditions in a partial nitration system. *Water Research* 45: 5934–5944.

Liu, Y., Ni, B.-J., Law, Y., Byers, C., Yuan, Z. 2014. A novel methodology to quantify nitrous oxide emissions from full-scale wastewater treatment systems with surface aerators. *Water Research* 48: 257–268.

Mao, J., Jiang, X.Q., Yang, L.Z., Zhang, J., Qiao, Q.J., He C.D., Yin, S.X. 2006. Nitrous Oxide Production in a Sequence Batch Reactor Wastewater Treatment System Using Synthetic Wastewater. *Pedosphere* 16(4): 451–456.

Mazurkiewicz, M. 2012. Usuwanie związków azotu ze ścieków w oczyszczalni w Kostrzynie nad Odrą. *Inżynieria środowiska* 127(27): 5–15.

Miąsik, M., Czarnota, J., Tomaszek, J.A. 2013. Emisja gazów cieplarnianych z obiektów oczyszczalni ścieków. *Czasopismo Inżynierii Lądowej, Środowiska i Architektury* 3(60): 253–264.

Miksch, K., Sikora, J. 2010. Biotechnologia ścieków. Warszawa: Wydawnictwo Naukowe PWN.

Montzka, S.A., Reimann, S., Engel, A., Kruger, K., O'Doherty, S., Sturges, W.T., Blake, D., Dorf, M., Fraser, P., Froidevaux, L., Jucks, K., Kreher, K., Kurylo, M.J., Mellouki, A., Miller, J., Nielsen, O.-J., Orkin, V.L., Prinn, R.G., Rhew, R., Santee, M.L., Stohl, A., Verdonik, D. 2011. Ozonedepleting Substances (ODSs) and Related Chemicals, Chapter 1 in Scientific Assessment of Ozone Depletion. 2010. Global Ozone Research and Monitoring Project, *Report* No. 52. World Meteorological Organization, Geneva, Switzerland, 516 pp.

Mosier, A.R. 1994. Nitrous oxide emissions from agricultural soils. *Fertilizer Research* 37: 191–200.

Mosier, A.R., Duxbury, J.M., Freney, J.R., Heinemeyer, O., Minami, K. 1996. Nitrous oxide emissions from agricultural fields: Assessment, measurement and mitigation. *Plant and Soil* 181: 95–108.

Mosier, A.R., Kroeze, C., Nevison, C., Oenema, O., Seitzinger, S., van Cleemput, O. 1998. Closing the global atmospheric N_2O budget: Nitrous oxide emissions through the agricultural nitrogen cycle. *Nutrient Cycling in Agroecosystems* 52: 225–248.

Mosier, A., Kroeze, C., Nevison, C., Oenema, O., Seitzinger, S., van Cleemput, O. 1999. An overview of the revised 1996 IPCC guidelines for national greenhouse gas inventory methodology for nitrous oxide from agriculture. *Environmental Science and Policy* 2(3), 1999: 325–333.

Nyćkowiak, J., Leśny, J., Olejnik, J. 2012. Ocena bezpośredniej emisji N_2O z gleb użytkowanych rolniczo województwa wielkopolskiego w latach 1960–2009 według metodologii IPCC, *Water-Environment-Rural Areas* 4(40): 203–215.

Osada, T., Kuroda, K., Yonaga, M. 1995. Reducing nitrous oxide gas emissions from fill-and-draw type activated sludge process. *Water Research* 29(6): 1607–1608.

Otte, S., Grobben, N.G., Robertson, L.A., Jetten, M.S.M., Kuenen, J.G. 1996. Nitrous oxide production by Alcaligenes faecalis under transient and dynamic aerobic and anaerobic conditions. *Applied and Environmental Microbiology* 62(7): 2421–2426.

Park, K.Y., Inamori, Y., Mizuochi, M., Ahn, K.H. 2000. Emission and control of nitrous oxide from a biological wastewater treatment system with intermittent aeration. *Journal of Bioscience and Bioengineering* 90(3): 247–252.

Park, H.-D., Wells, G.F., Bae, H., Criddle, C.S., Francis, C.A. 2006. Occurrence of ammonia-oxidizing Archaea in wastewater treatment plant bioreactors. *Applied and Environmental Microbiology* 72(8): 5643–5647.

Pattey, E., Edwards, G.C., Desjardins, R.L., Pennock, D.J., Smith, W., Grant, B., MacPherson, J.I. 2007. Tools for quantifying N_2O emissions from agroecosystems. *Agricultural and Forest Meteorology* 142: 103–119.

Podedworna, J., Żubrowska-Sudoł, M. 2004. Przegląd stanu wiedzy w zakresie biochemicznych przemian azotu w procesie oczyszczania ścieków. *Biotechnologia* 1(64): 127–141.

Prendez, M., Lara-Gonzalez, S. 2008. Application of strategies for sanitation management in wastewater treatment plants in order to control/reduce greenhouse gas emissions. *Journal of Environmental Management* 88: 658–664.

Quan, X., Zhang, M., Lawlor, P.G., Yang, Z., Zhan, X. 2012. Nitrous oxide emission and nutrient removal in aerobic granular sludge sequencing batch reactors. *Water Research* 46: 4981–4990.

Rajagopal, R., Beline, F. 2011. Nitrogen removal via nitrite pathway and the related nitrous oxide emission during piggery wastewater treatment. *Bioresource Technology* 102: 4042–4046.

Rapson, T.D., Dacres, H. 2014. Analytical techniques for measuring nitrous oxide. *Trends in Analytical Chemistry* 54: 65–74.

Rassamee, V., Sattayatewa, C., Pagilla, K., Chandran, K. 2011. Effect of oxic and anoxic conditions on nitrous oxide emissions from nitrification and denitrification processes. *Biotechnology and Bioengineering* 108(9): 2036–2045.

Rathnayake, R.M.L.D., Song, Y., Tumendelger, A., Oshiki, M., Ishii, S., Satoh, H., Toyoda, S., Yoshida, N., Okabe, S. 2013. Source identification of nitrous oxide on autotrophic partial nitrification in a granular sludge reaktor. *Water Research* 47: 7078–7086.

Ravishankara, A.R., Daniel, J.S., Portmann, R.W. 2009. Nitrous Oxide (N_2O): the dominant ozone-depleting substance emitted in the 21st century. *Science* 326(5949): 123–125.

Robertson, L.A., Kuenen, J.G. 1995. Nitrogen removal from wastewater. ECB6: Proceedings of the 6th European Congress on Biotechnology: 235–240.

Sadecka, Z. 2010. *Podstawy biologicznego oczyszczania ścieków*. Warszawa: Wyd. Seidel-Przywecki.

Schulthess, R., Kühni, M., Gujer, W. 1994. Release of nitric and nitrous oxide from denitryfying activated sludge. *Water Research* 29(1): 215–226.

87

Snip, L. 2010. Quantifying the greenhouse gas emissions of waste water treatment plants. *Environmental Sciences Netherlands*: 8–13.

Sun, S., Cheng, X., Sun, D. 2013. Emission of N_2O from a full-scale sequencing batch reactor wastewater treatment plant: Characteristics and influencing factors. *International Biodeterioration & Biodegradation* 85: 545–549.

Sümer, E., Weiske, A., Benckiser, G., Ottow, J.C.G. 1995. Influence of environmental conditions on the amount of N_2O released from activated sludge in a domestic waste water treatment plant. *Cell. Mol. Life Sci.* 51: 419–422.

Szewczyk, K.W. 2005. *Biologiczne metody usuwania związków azotu ze ścieków*. Warszawa: Oficyna Wydawnicza Politechniki Warszawskiej.

Tallec, G., Garnier, J., Billen, G., Gousailles, M. 2008. Nitrous oxide emissions from denitrifying activated sludge of urban wastewater treatment plants, under anoxia and low oxygenation. *Bioresource Technology* 99: 2200–2209.

Tallec, G., Garnier, J., Billen, G., Gousailles, M. 2006. Nitrous oxide emissions from secondary activated sludge in nitrifying conditions of urban wastewater treatment plants: effect of oxygenation level. *Water Research* 40(15): 2972–2980.

Thoern, M., Soerensson, F. 1996. Variation of nitrous oxide formation in the denitrification basin in a wastewater treatment plant with nitrogen removal. *Water Research* 30(6): 1543–1547.

Thomsen, M., Lyck, E. 2008. Emission of CH_4 and N_2O from Wastewater Treatment Plants (6B). *NERI Technical Note* no. 208, Denmark.

Tomaszek, J.A., Gruca-Rokosz, R., Koszelnik, P. 2008. Stabilne izotopy. Ważne narzędzie w inżynierii środowiska. *Przemysł chemiczny* 87/5: 593–595.

Turner, D.A., Chen, D., Galbally, I.E., Leuning, R., Edis, R.B., Li, Y., Kelly, K., Phillips, F. 2008. Spatial variability of nitrous oxide emissions from an Australian irrigated dairy pasture, *Plant Soil* 309: 77–88.

United States Environmental Protection Agency. 2006. Global anthropogenic non-CO_2 greenhouse gas emissions: 1990 to 2020. Washington, DC: US-EPA.

USEPA. 2009. Inventory of U.S. Greenhouse Gas Emissions and Sinks: 1990–2007. U.S. Environmental Protection Agency, Washington DC. *Report* No. EPA 430-R09-004.

Wang, Ch., Zhu, G., Wang, Y., Wang, S., Yin, Ch. 2012. Nitrous oxide reductase gene (nosZ) and N_2O reduction along the litoral gradient of a eutrophic freshwater lake. *Journal of Environmental Studies* 25(1): 44–52.

WMO, 2009a. The State of Greenhouse Gases in the Atmosphere Using Global Observations through 2008, *World Meteorological Organization Greenhouse Gas Bulletin*, 5:23. [Dostęp 14.02.2014] http://www.wmo.int/pages/prog/arep/gaw/ghg/documents/ghg-bulletin2008_en.pdf

WMO, 2009b. GAW Report No. 185 Guidelines for the Measurement of Methane and Nitrous Oxide and their Quality Assurance.

Wuebbles, D.J. 2009. Nitrous oxide: no laughing matter. *Science* 326: 56–57.

Wunderlin, P., Mohn, J., Joss, A., Emmenegger, L., Siegrist, H. 2012. Mechanisms of N_2O production in biological wastewater treatment under nitrifying and denitrifying conditions. *Water Research* 46: 1027–1037.

van Niel, E.W.J., Braber, K.J., Robertson, L.A., Kuenen, J.G. 1992. Heterotrophic nitrification and aerobic denitrification in Alcaligenes faecalis strain TUD. *Antonie Van Leeuwenhoek* 62: 231–237.

von Schulthess, R., Wild, D. & Gujer, W. 1994. Nitric and nitrous oxide from denitrifying activated sludge at low oxygen concentrations. *Water Sci. Technol.* 30: 123–132

Quan, X., Zhang, M., Lawlor, P.G., Yang, Z., Zhan, X. 2012. Nitrous oxide emission and nutrient removal in aerobic granular sludge sequencing batch reactors. *Water Research* 46: 4981–4990.

Yu, R., Kampschreur, M.J., van Loosdrecht, M.C.M., Chandran, K. 2010. Mechanisms and specific directionality of autotrophic nitrous oxide and nitric oxide generation during transient anoxia. *Environmental Science & Technology* 44(4): 1313–1319.

Zhang, M., Lawlor, P.G., Li, J., Zhan, X. 2012. Characteristics of nitrous oxide (N_2O) emissions from intermittently-aerated sequencing batch reactors treating the separated liquid fraction of anaerobically digested pig manure. *Water, Air and Soil Pollut* 223: 1973–1981.

Zheng, H., Hanaki, K., Matsuo, T. 1994. Production of nitrous oxide gas during nitrification of wasterwater. *Water Science and Technology* 30(6), 133–141.

Progress in Environmental Engineering – Tomaszek & Koszelnik (eds)
© 2015 Taylor & Francis Group, London, ISBN: 978-1-138-02799-2

Author index

Printed and bound by CPI Group (UK) Ltd, Croydon, CR0 4YY

18/10/2024

01776219-0017